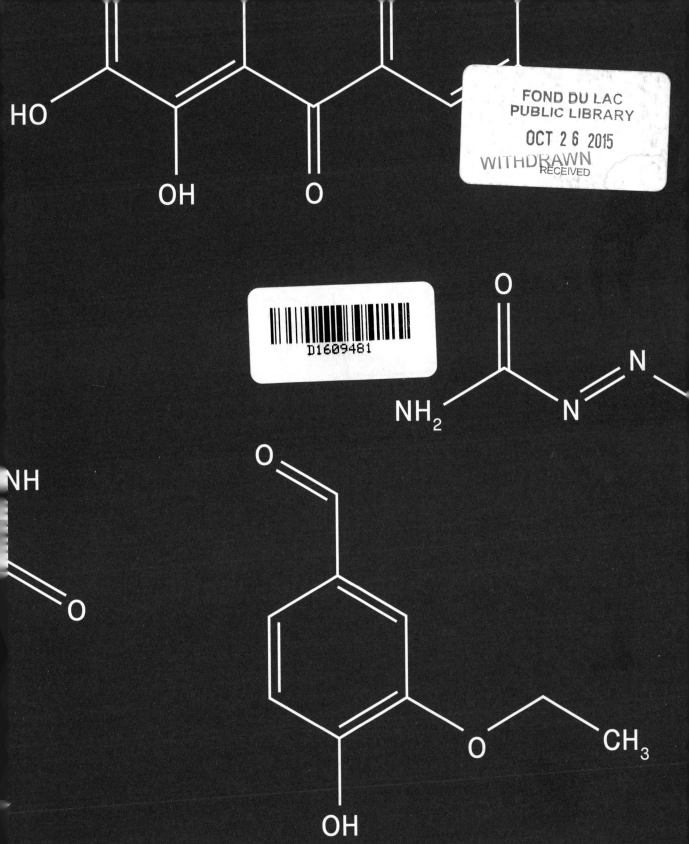

INGREDIENTS:

**A Visual Exploration of
75 Additives & 25 Food Products**

Dwight Eschliman
Text by Steve Ettlinger

Regan Arts.

65 Bleecker Street
New York, NY 10012

Photography Copyright © 2015 by Dwight Eschliman
Text © 2015 by Steve Ettlinger

All rights reserved, including the right to reproduce this book or portions thereof in any form whatsoever. For information address Regan Arts Subsidiary Rights Department, 65 Bleecker Street, New York, NY 10012.

First Regan Arts hardcover edition, September 2015.

Library of Congress Control Number: 2015930629

ISBN 978-1-941393-31-4

Book design by Manual
Jacket art by Dwight Eschliman
Additive and Food styling by Vanessa Chu

Synonyms, functions, molecular formulas, and molecular structures courtesy of Alyson E. Mitchell, PhD, Univ. of California Davis

Printed in China

10 9 8 7 6 5 4 3 2 1

Throughout the book the authors discuss readily available consumer products by their generally accepted brand names. Some of these names may also be trademarks. The uses of these brand names are for editorial purposes only and their use in this book does not constitute an endorsement or authorization from the owner of the mark.

Chart on pages 250-251 is based on information contained in FDA Publication No. 79—2115 and on www.fda.gov/Food/IngredientsPackagingLabeling/FoodAdditivesIngredients/ucm094211#types.

Portions of this text previously appeared in or are based on small portions of Steve Ettlinger's book *Twinkie, Deconstructed* (Hudson Street Press, 2007; Plume, 2008) and are included here with permission of the author and the publisher.

Key to front and back cover:

Front cover

Top row, left to right:
Caffeine
Riboflavin
Soy protein

Middle row, left to right:
Xanthan gum
Wheat grass
Poppy seeds

Bottom row, left to right:
Blue No. 1
Artificial flavor
Succinic acid

Back cover

Top row, left to right:
Beta-carotene
Wheat
Sodium triphosphate

Middle row, left to right:
Carbonated water
Folic acid
Ferrous fumarate

Bottom row, left to right:
Soy lecithin
Diglycerides
Chlorophyll

This book is for anyone wondering what's in their food.

Contents

viii—xi
Introduction

**Part 1—
75 Additives**

02—03
Acesulfame potassium

04—05
Agar

06—07
Alginate

08—09
Annatto

10—11
Anthocyanins

12—13
Ascorbic acid

14—15
Autolyzed yeast

16—17
Azodicarbonamide (ADA)

18—19
Baking soda

20—23
Beta-carotene

24—25
BHA and BHT

26—27
Caffeine

28—31
Calcium sulfate

32—33
Caramel color

34—35
Carmine

36—39
Carrageenan

40—43
Caseinate
(calcium and sodium)

44—45
Cellulose gum

46—47
Chlorophyll

48—49
Citric acid

52—53
Corn Ingredients

54—55
Corn flour

56—57
Cornstarch

58—59
Modified cornstarch

60—61
Maltodextrin

62—63
Dextrose

64—65
Corn syrup

66—69
High-fructose corn syrup

70—73
Sorbitol and mannitol

76—77
Diacetyl

78—79
Disodium inosinate

80—81
Ethylenedi-
aminetetraacetic acid
(EDTA)

84—85
Ethyl vanillin

86—87
Ferrous sulfate

88—89
Folic acid

90—93
Gelatin

94—95
Gum arabic

96—97
Isoamyl acetate

98—99
Lactic acid

100—101
Lycopene

102—103
Milk thistle extract

104—105
Mono- and diglycerides

106—107
Monocalcium phosphate

108—111
Monosodium glutamate
(MSG)

112—113
Neohesperidin
dihydrochalcone (NHDC)

114—115
Niacin

116—117
Phosphoric acid

120—121
Polydimethylsiloxane
(PDMS)

122—125
Polyglycerol
polyricinoleate (PGPR)

126—127
Polysorbate 60

128—129
Potassium chloride

130—133
Probiotics

134—135
Propionate (calcium and sodium)

136—139
Propylene glycol

140—141
Red No. 40 and Yellow No. 5

142—143
Riboflavin

144—147
Salt

148—149
Shellac

150—151
Silicon dioxide

152—153
Sodium acid pyrophosphate (SAPP)

154—155
Sodium benzoate

156—157
Sodium nitrite

160—161
Sodium stearoyl lactylate (SSL)

162—163
Sorbic acid

164—165
Soy Ingredients

166—167
Partially hydrogenated vegetable oil

168—169
Soy lecithin

170—171
Soy protein isolate (SPI)

174—175
Stevia

176—177
Sucralose

178—181
Sugar

182—185
Sweet dairy whey

186—187
Taurine

188—189
Tertiary-butylhydroquinone (TBHQ)

190—191
Thiamine mononitrate

192—193
Titanium dioxide

194—195
Xanthan gum

Part 2—25 Processed Food Products

200—201
Amy's Burrito Especial

202—203
Campbell's Chunky Classic Chicken Noodle Soup

204—205
Doritos Cool Ranch Flavored Tortilla Chips

206—207
Dr Pepper

208—209
General Mills Raisin Nut Bran

210—211
Hebrew National Beef Franks

212—213
Heinz Tomato Ketchup

214—215
Hidden Valley The Original Ranch Light Dressing

216—217
Hostess Twinkies

218—219
Klondike Reese's Ice Cream Bars

220—221
Kraft Cool Whip Original

222—223
Kraft Singles - American Skim Milk Fat Free

224—225
McDonald's Chicken McNuggets

226—227
MorningStar Farms Original Sausage Patties

228—229
Nabisco Wheat Thins

230—231
Naked Green Machine 100% Juice Smoothie

232—233
Nestlé Coffee-Mate Fat Free The Original Coffee Creamer

234—235
Ocean Spray Cran-Grape Juice Drink

236—237
Oroweat Healthy Multi-Grain Bread

238—239
PowerBar Performance Energy Bar Oatmeal Raisin

240—241
Quaker Instant Oatmeal Strawberries & Cream

242—243
Red Bull Energy Drink

244—245
Snickers Bar

246—247
Trident Perfect Peppermint Sugar Free Gum

248—249
Vlasic Ovals Hamburger Dill Chips

250—251
Food Additive Functions

252—253
Glossary

255
Acknowledgments

Introduction

Almost everyone eats processed food. Even the most devoted pure-food fan smokes, pickles, freezes, or cooks food—the most basic forms of preserving food that have been with us since the dawn of mankind. However, many of us are more curious than ever about the food ingredients that don't seem like real food: the additives (those things with the long, unfamiliar, and complex chemical names listed on food packages). This interest may have something to do with the fact that additives are mostly not real food. They also catch our attention, in part, because we don't know what the heck the names mean and, in part, because, let's face it, they simply seem scary.

We hope to address that uneasiness with this unique book. Dwight and I are often amazed by items we find among the daily elements of our lives, and love learning about them. We have some other things in common, too, like our insatiable desire to know what's actually in our food.

Raised on homemade wheat-gluten patties and scrambled tofu colored with turmeric to resemble eggs, processed foods were out of Dwight's reach for the first part of his life. After trading the safety of his mother's kitchen for a college cafeteria, he loosened up a bit—not so much that he substituted all of his fruits and vegetables with processed foods, but enough that whole foods shared the plate with many of the usual processed suspects of the standard American diet. Once he became a father, he began to think about food analytically for the first time. As a successful advertising and editorial photographer, Dwight took his inspiration for food deconstruction from conversations about food ingredients on photo sets with an inspired chef and a talented stylist. He sought to combine his curiosity about food with his obsessive-compulsive nature, which eventually led him to me.

I have worked as an assistant chef, lived in Paris, where I found myself eating at top restaurants as part of my work, and have written several books on food-related subjects. As I researched the origins of various foods, especially the ingredients in ethnic cuisines and beer, I began wondering about artificial food ingredients and took up the habit of reading labels very closely. When my young children, who mostly ate whole foods, started quizzing me about the processed-food labels I was staring at, I realized that the way to do an in-depth investigation on the subject was to deconstruct the ingredient label of a well-known processed-food product. I soon settled on Hostess® Twinkies® and wrote a book, *Twinkie, Deconstructed*, about where all the Twinkies ingredients come from, how they are made, and what they do. I like digging into the details behind complex but common things. Instead of taking photos, though, I took trips to the mines, factories, and labs to see for myself the sources of such food additives.

One day I saw an enormous spike in my Web site traffic, which I found was linked to Dwight's self-published project, *37 or So Ingredients*, a collection of photographs of Twinkies ingredients that had gone viral. Thus art, science, and the loves of wonderment and deconstruction inspired Dwight and me to join together to create this visual representation of the food additives that are found in the ingredient lists of common processed foods.

About Food Additives

Anyone who cooks knows that you must almost always process food somehow to eat it. Unless you are a grazing animal, at a minimum, you clean it or you cook it. But industrial, mass-market, commercial food producers add things (additives) to food (basic foodstuffs) or even packaging for at least one of four reasons: to make the food product more nutritious, to make it easier to prepare, to make it more appealing, or to make it stay fresh longer.

The ingredients in our book are additives that fill at least one of these functions; some fill several (the Food and Drug Administration lists seventeen subsets of functions [p. 250]). Some replace foods that are prone to spoilage, such as fresh eggs. Others, such as various corn syrups, replace foods that might be more expensive, such as sugar. Some, such as the vitamins in enriched flour, replace the natural ingredients that were removed in processing. Others simply add a healthy one. And many are there to somehow extend a product's shelf life, the holy grail of food product manufacturers. They all are part of the creation of what consumers perceive as inexpensive, convenient foods.

When you bake a cake at home, you don't use most of the additives shown here. You might mix a batter with more elbow grease for one recipe than another; you might refrigerate or warm up various ingredients (such as butter) to make them blend more easily. You might cover a stored item tightly to make it last longer in the fridge without drying out or spoiling. You might form the top so it looks nicer. You might buy a special cupcake transporter box to protect your goods. But you don't need any additives to do these things.

On the other hand, when you bake a cake or make some commercial food product by the millions in a large factory with industrial machinery and ship it around the country, where it sits on store shelves for weeks, you might add something to a batter to make it easier to pump through hoses. You might add something to keep the bubbles in a batter from getting crushed at the bottom of an enormous kettle. You might add something to keep the final product from losing moisture or flavor in storage or so it doesn't collapse during transit. You might add something or use special ingredients so it doesn't spoil quickly. In short, you use food additives to achieve the scaled-up goals that the home cook addresses quite differently.

Since World War II, the availability and use of food additives have evolved considerably, especially recently. The war created demand for chemical research; the postwar economy created demand for convenience foods. These days, consumer demands tend to focus on carefully created food products that deliver certain health benefits—so-called functional foods such as high-fiber, low-fat, and no-sugar products. The result is food products filled with more and more additives and much longer and more complex ingredient lists. It can get quite confusing.

About Food Product Labels

Food additives come with a lot of baggage. Related terms are often obtuse. This stems from the fact that their names are a mix of scientific and cultural traditions, that they must be approved by a variety of outfits, and that they are regulated by a variety of rules and laws across borders. Food additives receive approval only after years of testing and correspondence with the authorities (dietary supplements are not FDA-regulated), but some considered safe in one country are banned in another. Some of that irregularity stems from cultural and political differences and some comes from the relative influence of the food industry. Research is vetted by disparate groups, including the Food and Agriculture Organization of the United Nations, the World Health Organization, and the Codex Alimentarious Commission.

Hidden from the common shopper are regulations concerning allowable amounts of an additive in a food product. Some are restricted to a minuscule parts per million (ppm), while others, having been in use for years, earn the vague classification of GRAS (Generally Recognized as Safe) and are limited to doses defined even more vaguely as "good manufacturing practices."

Understanding the ingredient section on a food label is otherwise fairly simple. The FDA requires that common, not scientific, names be used. The ingredients are listed in the descending order of their presence by weight, but no other information as to proportion is offered. That means your first two items might be 95 percent of the product, such as the flour and sweeteners in snack cakes (it is up to you to add the sugar and corn syrups together for the total sugars present). Obviously, the items listed after "2% or less" are there in minuscule proportions. The manufacturer's secret recipe is kept safe.

Two of the ingredients in this book—baking soda and monocalcium phosphate—are part of the history behind food labels. They are also cultural markers of sorts, at least for Americans. They were first mixed together and sold as baking powder back in 1859, creating one of the world's first mass-marketed convenience-food ingredients. As consumer demand grew in the late 1800s, marketing efforts intensified, leading to wild and injurious claims that some baking powder brands were poisonous. This, along with some well-publicized food-safety problems, led to a political effort for honesty in packaging, eventually inspiring the first food-labeling law, the Federal Food and Drugs Act of 1906 (which morphed into the FDA). Regulations have evolved continuously ever since.

Where We're Coming From
Some readers might expect a firm indictment of artificial food ingredients, but they will not find that in this book. This is a visual exploration with a popular-science angle, not a polemic. We are not here to tell you that artificial ingredients are bad for you, or what to eat (other writers have amply covered those issues, and besides, everyone should know to eat mostly fruit, vegetables, and whole grains). We are simply curious about these ingredients and assume that many of you are too. We ask, "What does it look like?" and "Why do they put this in my food?" This book will make you think about food additives as real stuff, not just some strange words on a label. (Still, we freely admit that our diets are oriented toward apples and broccoli rather than foods that require a long list of additives.)

Approved additives have been so extensively tested by the FDA that it is safe to assume that they are OK for consumption. However, there's always some risk. Some watchdog groups suggest avoiding certain additives or campaign for more testing. After all, many food additives have been banned or discovered to be unhealthy after years of use. Partially hydrogenated vegetable oil, better known as shortening (think Crisco and margarine), which was one of the very first artificial ingredients and has been around since 1911, was recently revealed to contain harmful trans fats (Crisco is now trans fat–free). If you are concerned about unsafe additives, especially due to allergies or other health concerns, you need to find well-vetted, professional scientific information as to what works for you.

One thing we clearly oppose is chemophobia. The idea that everything with a long chemical name is definitely bad for you is just plain stupid, but that does not stop some people, such as the occasional frenzied blogger, from freaking out over certain food additives. While it might be a good idea to avoid specific artificial ingredients, it is an even better idea to do so knowingly (sugar and salt may do more harm than artificial ingredients). Otherwise you might find yourself in a situation where you are worrying about anything with dihydrogen monoxide in it, because dihydrogen monoxide is a main component of all kinds of nasty, toxic things, like paint, sulfuric acid drain cleaner, and acid rain. It is actually fatal if you accidently inhale it! It is also fatal in large doses! It is used to cool nuclear power plants! One might panic and ask why on earth you would want something like that in your food! Well, FYI, dihydrogen monoxide is often identified by its chemical formula, H_2O. Yes—plain old water. We prefer science rather than blather to quell our fears. Panic is not pretty.

In any case, the debate is not really about artificial ingredients being good or bad for you. The real debate concerns whether a diet filled with highly processed foods is as likely to be as good for you as one that minimizes them, and whether a food and agriculture system that relies extensively on artificial food ingredients, petroleum, and corporate, monoculture farming is sustainable and a good political and environmental policy. We don't get into that because, simply put, we're dealing here with art and science, not policy.

Dwight has photographed samples of additives as you might buy them (some are available to food processing companies in a variety of forms, such as powdered, granulated, or liquefied). After interviewing scientists and industry professionals, and after consulting books and industry websites, I have explained in as few words as possible where these things are made, how they are or came to be made, and what they are used for. Consider this a mere appetizer of information.

There are thousands of food additives; we are showing you only 75 of the most commonly used additives, chosen to represent a range of types and functions. In short, these are the kinds of additives in the food that any one of us might be buying or consuming every day.

If food ingredient labels make your eyes glaze over, we hope that this book will open them instead. We hope that this little bit of art and science will make your grocery shopping more informed and, ultimately, more enjoyable. Bon appétit!

—Steve Ettlinger

Part 1

75 Additives

Acesulfame potassium

$C_4H_4KNO_4S$

CFR number
172.800

E number
E950

CAS number
55589-62-3

Synonyms or siblings
Acesulfame K, Ace K, Acesulfame potassium

Function
Appeal-Sweetener

Description

Back in 1967, at the end of what was probably a long lab session in Germany, Hoechst AG chemist Karl Clauss absentmindedly licked his finger. Normally this is not a good thing to do in a chemistry lab. However, in a long and hallowed tradition of random luck leading to big scientific discoveries, he was startled to find that his finger tasted supersweet. Like, 200-times-the-sweetness-of-sugar sugar sweet. He had accidentally discovered acesulfame potassium, a calorie-free artificial sweetener.

Now marketed to consumers as Sunett, this sweetener is made by a complex organic chemical reaction that starts with petrochemicals processed to make acetic acid, which is then further processed to create acetoacetic acid. In a Celanese-owned plant in Frankfurt, Germany, the acetoacetic acid is combined with potassium (chemical symbol: K) to make the sweet white crystals.

One of the newer artificial sweeteners, acesulfame potassium was approved by the FDA only in 1988, but it is now found in thousands of food products. Also known as acesulfame K, it is usually blended with other sweeteners (such as aspartame or sucralose [p. 176]) in order to mask its bitter aftertaste. Amazingly, this blend of ultrasweet sweeteners is more powerful together than the individual chemicals are on their own.

An advantage of acesulfame potassium is that it is stable enough to keep its sweetness despite exposure to high heat, so it is popular in baked goods, candies, and desserts, as well as beverages and tabletop sweeteners. It is also useful in the manufacture of various pharmaceuticals, including cough drops and syrups. Acesulfame potassium is stable in another way, too: it passes through your body completely unchanged, a vertitable sweet nothing.

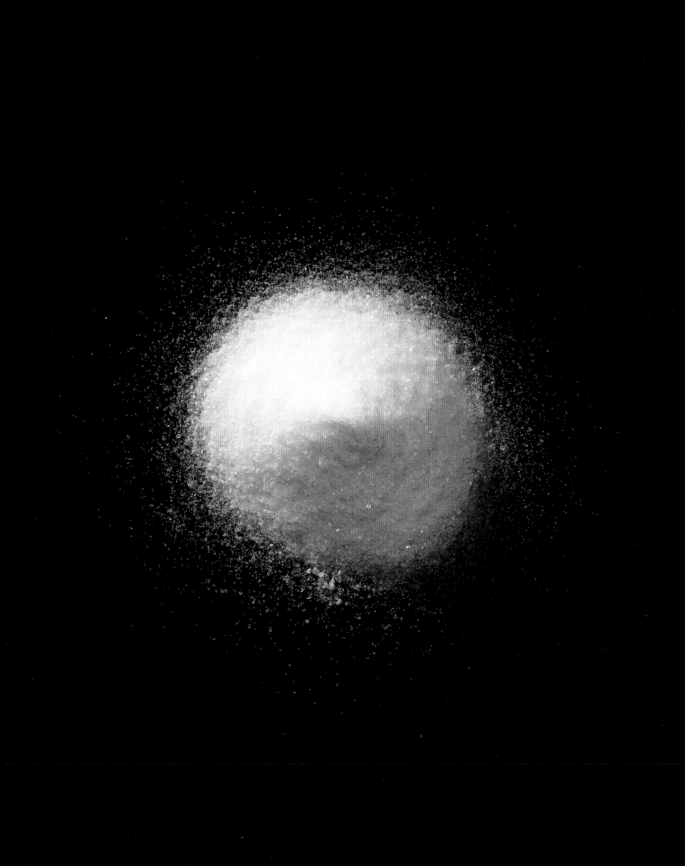

Agar

CFR number
184.1115

E number
E406

CAS number
9002-18-0

Synonyms or siblings
Agar-agar, Agar-agar flake, Agar-agar gum, Polysaccharide complex

Function
Process and Prep-
Thickening Agent

$(C_{12}H_{18}O_9)n$
Agarose polymer

Description

Agar is nothing more than processed wild red seaweed, a red algae (species Gelidium and Gracilaria) harvested mostly off the coasts of Spain, Portugal, and Morocco, as well as in small quantities from Japan, Ireland, and off the coast of San Diego, California. It has a centuries-long track record in Japan, which until World War II was the source of most agar, and China, where some say a 16th-century Chinese emperor first noted agar's gelling abilities.

The actual workhorse in this common plant is an unbranched polysaccharide that lines its cells. Known by many as a vegetarian alternative to gelatin [p. 90], it finds its way into puddings, custards, jelly candies, low-calorie jams, canned soups, ice cream, doughnuts, and, of course, gelatin fruit desserts. It is so useful that many call it "the queen of gelling agents." Others call it agar-agar, gelose, kanten (in Japanese), Japanese or Chinese isinglass (which is actually incorrect; isinglass is made from fish bladders, not seaweed), and simply vegetable gelatin. Agar (as with other seaweeds) is often processed by freezing or compressing it until it becomes a dry block that is then ground into powder.

Agar is 80 percent fiber-rich, so it is sold as a health remedy and a fad diet product said to absorb water in your stomach and make you feel more full (it's also recommended as a laxative, so filling up on agar may be a bad choice). Fans of gelatinous and gooey Filipino desserts such as sago, red and black gulaman, buko pandan, and more are essentially fans of agar. So are lovers of the Vietnamese dessert thach and the Burmese sweet jelly kyauk kyaw. Agar is gluten-free, fat-free, and low in sodium. No wonder you can find it for sale at health food stores.

It's ironic that agar is a popular ingredient in both desserts and dentists' offices. It's used to make dental impressions used for fillings and other corrective procedures. However, it is also the material of choice for making salt bridges in electrochemistry, clarifying beer, sizing paper and fabrics, and supplying nutrition and lodging in ant farms. Russians use it to replace pectin in jellies. Most of us know its scientific use from high school biology, where agar plates—petri dishes with a thin layer of agar on the bottom—are used for growing microorganisms of all sorts. That's why you'll also find it for sale online among high school science fair supplies.

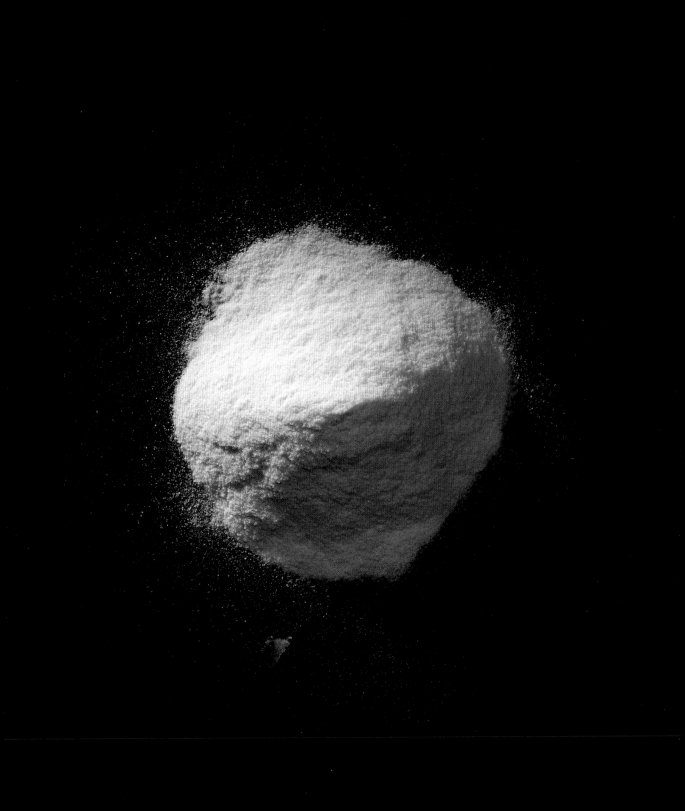

Alginate

CFR number
184.1724
(Sodium alginate)
184.1011
(Alginic acid)

E number
E400

CAS number
95328-14-6

Synonyms or siblings
Algin, Alginic acid, Potassium alginate

Function
Process and Prep-Stabilizer, Thickening Agent

$(C_6H_7O_6Na)n$
Sodium alginate

Description

Alginate is great for creating rubbery masks for movie monsters; forming dental impressions and body molds; covering and acting as a flux on welding rods; coating paper for better ink retention; making dyes work better in textile printing; and as a gelling, stabilizing, and thickening agent in foods. It also stabilizes the foam in a mass-produced beer head (microfilters remove the natural proteins that would otherwise hold it together). Alginate's wide range of viscosity is affected by temperature, acidity, and rate of stirring (i.e., aspects of a serious recipe), an attribute called either "non-Newtonion flow" or by the fetching name "pseudoplasticity."

Alginate is commonly found in thick salad dressings, fruit juices, fruit pie fillings, ice cream and sorbet (to prevent crystal formation and to delay melting), low-fat spreads, and flavored yogurts. It is also the go-to ingredient for what the food industry calls "restructured" or "reformed" foods, meaning it's used to bind bits of leftover meat and meat trimmings and other animal parts together to form more products that might include dog food, meatloaves, and food-like items such as some kinds of chicken nuggets.

And it all comes from seaweed.

This magical stuff is what lines the cell walls of brown seaweed. Even the most technologically advanced company starts its alginate process with freshly harvested giant seaweed (kelp) that is chopped, washed with acid, and dissolved in lye (sodium hydroxide) or some other alkali. Because the essence of the stuff is alginic acid, it is usually reacted with sodium (sodium carbonate) or calcium to make a neutral salt, such as sodium or calcium alginate. (If the recipe calls for it to remain an acid, such as is used in molecular cuisine, it is often mixed with propylene glycol [p. 136] and given the name propylene glycol alginate, or PGA.) The last processing steps are to simply dry it and then grind it all into a fine powder.

Although the first scientist to figure this all out was a British chemist in 1893, mass production didn't take off, as with so many processed food ingredients, until the end of World War II. There was a small exception: it was first used to line cans and to stabilize ice cream as early as 1934. One of its main attractions is that it can absorb more than 200 times its weight in water, and quickly. These days, in addition to the familiar uses listed above, it is a favorite of molecular gastronomy chefs for spherification when creating such un-nuggetlike items as olive oil caviar and carbonated mojito pods.

Annatto

CFR number
73.3000

Synonyms or siblings
Roucou, Achiote, Bixin, Bija

E number
E106b

Function
Appeal-Color

CAS number
8015-67-6

$C_{25}H_{30}O_4$
Bixin

Description

One food perfectly exemplifies what annatto looks like and is used for: orange cheddar cheese. Many consumers think cheddar cheese is naturally orange, which is not surprising since dairies have been coloring their cheddar since the 16th century, attempting to make inferior cheeses resemble the top-quality ones made yellowish by the high carotene content of fresh spring and summer milk. Likewise, some may think of Spanish rice as naturally yellow, when in fact it is annatto doing the job.

An added benefit to using annatto in cuisine is that it imparts a delicately earthy, peppery flavor to dishes. For this reason it is the essential ingredient in the popular Latin American condiment sazón, and is often called the "poor man's saffron."

The Latin name of the evergreen shrub it comes from is Bija or Bixa orellana. The species is named after Francisco de Orellana, the Spanish explorer who was the first European to travel the length of the Amazon River. The genus name Bixa may be derived from the Portuguese *bico*, for "beak," referring to the beaklike shape of the plant's seedpods. Often called the lipstick tree because of its use in cosmetics, Bixa produces large seedpods full of bright red-orange pulp, the source of the strong dye. Its dark red seeds are processed separately, often to color cooking oil or as a dye source, via water extraction. These seeds are especially popular in Filipino dishes, such as kare-kare, and are commonly sold with Hispanic seasonings simply as "annatto seeds" or "achiote" as well as "Bija" (in the Carribbean). Annatto costs slightly less than its more artificial alternatives.

The early Spanish explorers introduced annatto around the world. Now, it is produced in many countries; Brazil is in the lead, followed by Peru and Kenya. About half of the world's production goes into dye and about half stays in Brazil, Peru, and Ecuador in the form of seeds that are used as a spice or condiment.

While it is established in traditional culture and cuisine, annatto also plays an important role in modern products; these range from margarine (where it has been largely replaced by synthetic beta-carotene [p. 20], its less expensive, tasteless, red-orange cousin) to meats, sausage casing, snack foods, baked goods, seasonings, pastry, and smoked fish. The European Union has issued a limited and disputed ban on its use in sauces and seasonings (some people have a slight reaction to it, but the science behind that is not clear). Likewise, there is no scientific understanding of how annatto might function as a medicine, though claims assert its healing qualities with fevers and kidney diseases. For its fans, it is the dye that won't let you die.

Anthocyanins

CFR number
73.169
(Grape color extract)
73.170
(Grape skin extract)

E number
E163

CAS number
11029-12-2

Synonyms or siblings
Anthocyanins, Grape skin extract, Grape color extract, Enocianina

Function
Appeal-Color
Preservative-Antioxidant

$C_{15}H_{11}O_6$
Cyanidin

Description

If you want a natural, lovely, dark purple-red color source, stomp on some red grapes, make wine, and keep the skins. Soak them, extract the color with water, alcohol, or acetone solvents, then concentrate, dry, and powder it, and you've got anthocyanins. They're the color pigments in grape skins. If you need more than grapes for your source, try elderberries, blueberries, black currants, acai berries, purple sweet potatoes, red cabbage, purple eggplants, red radishes, and black/purple carrots. Anthocyanins are actually phytochemicals found in almost all plants (but not in algae and mosses) and lend red and blue color to stems, roots, and flowers, too (the term "anthocyanin" comes from the Greek for "blue flower"). Still, the eggplant and the red grape are the ones most loaded with these pigments, at least as far as common vegetables and fruits are concerned; the coating of the black soybean's seed may have the absolute highest concentration, but that's not as easily processed. In any case, these days the anthocyanins in your food are almost all manufactured in Europe and China.

Coloring of this sort is immensely popular in food products that don't naturally contain the color of the food that lends its name to the product (i.e, grape soda). Anocyanins artificially color flavored desserts and beverages, cereals, salad dressings, candies, and frostings. More recently, they have become the colorant of choice in specialty waters, teas, energy drinks, spirits, and especially fruit juice drinks—some of which contain little to no fruit. Anthocyanins skew toward red-purple in slightly acidic foods and dark blue in more alkaline foods. Because they are water-soluble, their hue can change if the pH in the food products increases from natural or consumer acidification, or when cooked in the alkaline environment of aluminum pans. They are also used as pH indicators in concentrated solutions.

High temperature, light, ultraviolet rays, and oxygen degrade these fragile anthocyanins, as do digestive juices. However, scientists are trying to figure out how to best extract them from plants in ways that conserve their attributes because there is a possibility that these pigments—which are phytochemicals, members of the flavenoid family—have antioxidants and other qualities that may reduce coronary heart disease, improve visual acuity, fight cancer, delay aging, reduce inflammation, retard diabetes, and ward off viruses. The inspiration is clear, but at present, whatever health benefits they may have as additives or supplements remain unknown and are probably minimal owing to the fact they readily decompose when digested and are rapidly excreted. However, one can always hope.

Ascorbic acid

$C_6H_8O_6$

CFR number
182.3013

Synonyms or siblings
Vitamin C

E number
E300

Function
Nutrient-Enriching Agent
Preservative-Antioxidant

CAS number
50-81-7

Description

While most of us know that ascorbic acid is the scientific name for the very common and well-known vitamin C, much about it is surprising. Its name, for starters. Most of us link vitamin C with good health, energy, or fighting colds. However, its primary function—as stated succinctly by the National Institutes of Health—is "to prevent and treat scurvy, a disease caused by lack of vitamin C in the body." Scientists call it "ascorbic acid," which stems from a blend of scientific vocabulary for "without" and Latin for "scurvy" (a-scorbutus). That's not exactly what most of us have in mind as we down our orange juice in the morning. More importantly, that's not at all why it is in so many processed foods.

In 1934, vitamin C became the first vitamin to be produced synthetically. Though the best natural source is raw fruits and vegetables (a cup of broccoli packs a whopping 135 percent of the recommended daily dosage, more than in a medium orange), the synthetic version is actually made from a form of corn syrup [p. 64]. All but one of the few large factories making it are in China (the exception is in Scotland). Annual world production is estimated at more than 120,000 tons. Though the biotechnological process is essentially secret, it probably involves hydrogenating glucose into sorbitol, which is then fermented into sorbose, a natural sugar. That is fermented by another microbe, undergoes another chemical reaction, and is finally dehydrated into a powder.

We need vitamin C for good health for a number of reasons, but it is usually put in food products for its antioxidant qualities. A natural preservative, it is most common in frozen peaches, flavoring oils, apple juice, candy, and canned mushrooms. It keeps vegetables and fruits from turning brown by interrupting an enzymatic activity. Though it's a potent antioxidant, it is water-soluble, so it won't work as an antioxidant for fats. Finally, it adds some nice tartness to soft drinks.

While it may not cure the common cold, it does have its place in a surprisingly wide array of industrial chores. It is one of a few reactive chemicals in photographic developers and fluorescent microscope imaging, it removes metal stains from fiberglass, and it's part of a raw plastic manufacturing process. That's a long way from curing scurvy.

Autolyzed yeast

CAS number
977046-75-5

Synonyms or siblings
Baker's yeast, Brewer's yeast, Yeast autolyzate, Yeast extract

Function
Appeal-Flavor Enhancer

Description

Marmite, that brown gooey spread beloved by Brits (along with its Australian and Swiss counterparts, Vegemite and Cenovis, respectively), is mostly autolyzed yeast: a yeast extract, tempered with a touch of salt and spices, some vitamins, and vegetable extracts for flavor. Most people take a little of it on buttered bread for a healthy snack (it is a good source of B-complex vitamins).

Food processors use yeast extract (which is what most of them call autolyzed yeast), similarly to how they would use monosodium glutamate [p. 108] as a flavor booster and to add the umami taste. This is because yeast extract is about 5 percent glutamic acid, the base of MSG, so it supplies the umami taste without having to list the much-debated glutamate on the ingredient label.

Autolyzed yeast is listed as "natural" or "natural flavor," despite being highly processed, because the processing is just a gentle salting and heating that causes the yeast's digestive enzymes to work on its own cells ("autolyzed" means "self-digested"). This removes the tough outer-cell walls; nothing is transformed. The cell contents, notably the amino acids (proteins) such as glutamate, are merely released. Centrifuged and then dried into a powder, the yeast extract, as it is now called, is still loaded with all the proteins, minerals, and vitamins of yeast. The same autolyzation process, biologically speaking, occurs when making bread (when you let the dough rest), making wine (fermenting "on lees" or "sur lie"), or brewing beer (leaving the must or wort to sit). Brewers and winemakers need to control this carefully, as negligence during this step could ruin their product. Glutamates are a big part of the taste of wine and beer.

Yeast has been known to the world ever since the Dutch scientist Antonie van Leeuwenhoek observed it around 1680. Louis Pasteur elevated our knowledge of it dramatically, and argued in the 1850s and '60s with the notable German chemist Justus von Liebig about fermentation, taking the position that fermentation required living organisms (a biological process), while Liebig argued that it was a chemical and mechanical process. As a result, other scientists investigated fermentation while both Pasteur and Liebig were each proven only partly correct. However, the increased attention advanced our understanding of enzymes. It was Liebig, however, who developed a way of extracting and concentrating yeast as a food product.

In fact, brewer's yeast is often the source of autolyzed yeast; most of the yeast in Marmite is sourced from the Bass Ale brewery in Burton-Upon-Trent, England, only two miles from the Marmite factory. Marmite has been made since 1902 and is totally ingrained in British culture. Not to be outdone, some claim that its Australian version, Vegemite, is as Australian as kangaroos.

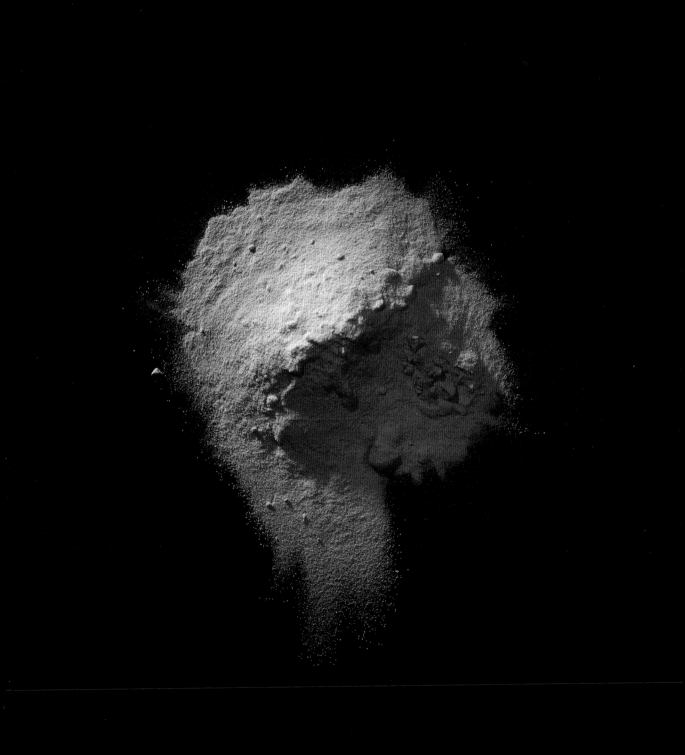

Azodicarbonamide (ADA)

$C_2H_4N_4O_2$

CFR number
172.8060

E number
E927

CAS number
123-77-3

Synonyms or siblings
Azobisformamide, Azodicarboxamide, Azodicaroxylic acid diamide

Function
Process and Prep-Emulsifier, Leavening Agent, Bleaching Agent

Description

Bakers prefer aged flour, and aging naturally takes time (and time equals money). To give the bakers what they want, flour mills mix a little oxidizing agent into the flour and—presto!—it turns white, rises higher, and ferments faster. This easy solution makes azodicarbonamide (ADA) a dough conditioner, dough strengthener, flour improver, and maturing agent that is so popular, it is found in hundreds of mass-produced breads and baked products.

Usually made from the chemicals urea and hydrazine in China, ADA has been on the processed food scene since 1962. If you see it on a label, it simply means that you are eating bleached flour that used ADA instead of potassium bromate (a similar ingredient that is a health risk if used in prohibited concentrations). Technically speaking, you are never actually eating ADA; once it hits the moisture in dough, it converts into harmless biurea, an organic water-insoluble chemical compound that passes through your body without effect. Furthermore, the legal limit of ADA is a minuscule 45ppm—the equivalent of 4.5 grams per 100 kilograms of wheat flour. That's pretty light.

ADA gives strength to dough and makes baked goods lighter because it helps with gas retention. It does this to plastics and rubber when it is used as a foaming agent to make bubbles for a controlled foam. You could say ADA puts the spring in your sandwich rolls and the bounce in your yoga mats. This fact once caught the attention of some publicity-minded but scientifically challenged bloggers who demanded that the Subway sandwich chain and others remove ADA from their breads, which they did. While the reaction of the food-marketing folks was understandable, it left the scientific and professional community scratching its collective head. The complainers themselves cited sources that confirmed it is safe in food and dangerous only to the factory workers making it—dangerous only in ways that most dusts are dangerous. Scientists were also stunned because so many food additives have industrial uses as well (including, of course, water). The whole thing boiled down to an occupational safety issue with no evidence of nutritional danger. While ADA remains banned in Europe and Australia, mostly due to its use in plastics, the FDA gives it the most nonrestrictive rating.

As for that fearsome-looking name, it merits explanation. "Azodicarbonamide" breaks down fairly easily into organic chemical science talk for the molecule's structure. "Azo" means that groups of nitrogen atoms with double bonds are attached to the molecule ("azo" comes from the French for nitrogen). "Dicarbon" means that there are two carbon atoms. "Amide" relates again to the nitrogen placement. It's a name only a chemist could love.

Baking soda

NaHCO$_3$

CFR number
184.1736

E number
E500

CAS number
144-55-8

Synonyms or siblings
Sodium bicarbonate, Bicarbonate of soda, Sodium acid carbonate, Sodium hydrogen carbonate

Function
Process and Prep-Leavening Agent

Description

Baking soda, also known as bicarbonate of soda, sodium bicarbonate, or simply bicarb, is familiar to almost everyone as an odor absorber, an effective antacid, a domestic cleaning product, and a baking ingredient. Along with vinegar, it's the (secret!) ingredient for actively foaming volcanoes in science fairs.

In fact, baking soda foams when it's mixed with any mild acid, whether it's in the form of a liquid like milk or lemon juice, or as a dry mix like monocalcium phosphate [p. 106]. The only difference is that the dry acid component reacts with baking soda once the mixture gets wet; this is the case in baking powder, in which baking soda is typically the base, or alkaline. When the alkaline and acid dry mixture hit water, they release a gas: plain old CO_2. The bubbles created get trapped in the batter as it solidifies into a cake or bread in the oven. The overall effect is to lighten the baked good, which is why it is called "leavening"; more specifically it's called "chemical leavening" to differentiate it from "natural leavening," which generally means yeast.

It's strange to think that the ingredient that makes cake fluffy and light comes from a rock. Specifically, domestically produced baking soda comes from a sedimentary rock called trona. In 1938, enormous concentrations were found in Green River, Wyoming; now, a handful of companies mine about 15 million tons of it each year. Trona is almost pure sodium carbonate (Na_2CO_3), or soda ash, which is one of the most important industrial chemicals in the world. It's used to make glass, paper, and soap, or a sodium source for other artificial food ingredients.

From ancient times to the 1800s, soda ash was made by burning plants or trees. The need for soda ash inspired widespread deforestation in New England when it was still a collection of British colonies. That changed in the 1860s, when Belgian chemist Ernest Solvay developed the Solvay process, a method that makes baking soda chemically from salt, ammonia, and limestone, and is now used throughout most of the world. Since the 1950s in the United States, however, baking soda has been made in Wyoming by bubbling some locally sourced CO_2 up through liquefied sodium carbonate in what is essentially a giant seltzer bottle. Once the baking soda is dried and processed to a uniform crystal size, most of it goes to cattle in feedlots. (Cows evolved to eat grass, and with their current unnatural grain diet—formulated to make them grow larger more quickly—they are in great need of antacid.) The rest goes to commercial bakeries or straight to grocery stores for our own use making bubbles in baked goods or reducing bubbles in our stomachs.

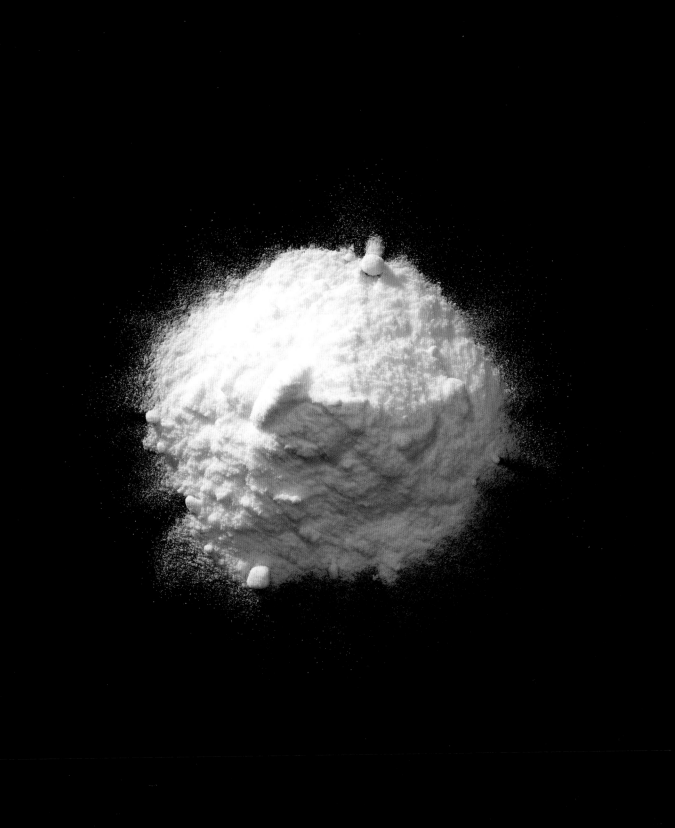

Beta-carotene

$C_{40}H_{56}$

CFR number
73.9500

E number
E160a

CAS number
7235-40-7

Synonyms or siblings
Betacarotene,
β-carotene,
CI food orange 5,
Provitamin A

Function
Appeal-Color
Nutrient-Fortifying Agent

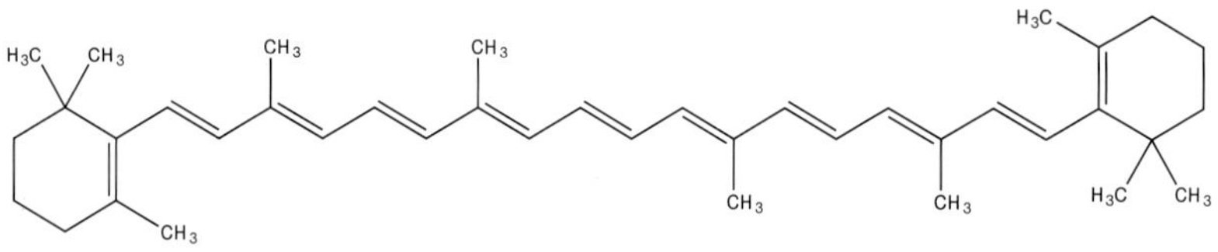

Description

Beta-carotene is the yellow crayon of the natural world, a popular colorant along with annatto [p. 8] and lycopene [p. 100]. It is the force behind the rich yellow of egg yolks, butter, and a significant part of the orange of carrots. A precursor to vitamin A—which the body turns into retinal for your eyes—beta-carotene is an antioxidant. It's good for you, unless you have too much of it, in which case you turn pale yellow-red with carotenodermia and could accidentally be diagnosed with jaundice. Alas, there is not even a recommended dietary allowance for carotenoids. According to years of research, beta-carotene supplements don't make up for a medical deficiency in vitamin A. Carotenoids are best digested as food, preferably lightly cooked.

Speaking of cooking, all carotenoids are fat-soluble, which means that you get more of their nutritional stuff when you chop the plants that contain them and cook them in oil or butter. Surprisingly, not all of these plants are yellow—spinach is an excellent source, even if its chlorophyll [p. 46] hides the yellow. Canned pumpkin is tops, as are sweet potatoes, cantaloupe, and, of course, carrots.

The industrial path to a reddish brown liquid that colors things bright yellow is similar to the process of soybean oil manufacturing [p. 164]. One of the main sources of this colorant is crude palm oil from Malaysian oil palm plantations, which have supplied companies since World War I. The crude oil is naturally bright orange-red because of

A German scientist first extracted beta-carotene from carrots and named it in 1831, but scientists have also figured carotenes to be among the oldest molecules around, found in organisms as long as 3 billion years ago. Carotenes have some serious history.

its high carotene content—the highest in any commonly found plant. It is mixed with a strong solvent, hexane, that leaches out the beta-carotene. Then the hexane is boiled off (and recycled), and the factory either mixes what's left with water and oil or dries it into a powder. This is a win-win for the manufacturer, as most oil clients prefer a clear-colored oil, and they end up producing a valuable colorant side product. Pure beta-carotene is also made via chemical synthesis in organic chemical or pharmaceutical plants, and by fermentation.

Beta-carotene, also written β-carotene, has been studied by scientists for close to 200 years, though much about the carotene family remains a mystery. A subset of carotenoids, these phytochemicals are naturally occurring plant pigments that are made mostly by photosynthesis. Ubiquitous chemicals in nature, there are about 600 known carotenoids. And yes, there is an alpha-carotene; the alpha and beta prefixes describe the location of rings of atoms in their molecules, and have nothing to do with a quality ranking system or age. A German scientist first extracted beta-carotene from carrots and named it in 1831, but scientists have also figured carotenes to be among the oldest molecules around, found in organisms as long ago as 3 billion years. Carotenes have some serious history.

BHA and BHT

$C_{11}H_{16}O_2$ and $C_{15}H_{24}O$

CFR number
182.3169 (BHA)
182.3173 (BHT)

E number
E320 (BHA)
E321 (BHT)

CAS number
25013-16-5 (BHA)
128-37-0 (BHT)

Synonyms or siblings
Butylated hydroxyanisole, tert-Butyl-4-hydroxyanisole, BOA (BHA)
Butylated hydroxytoluene, Butylhydroxytoluene, Bibutylated hydroxytoluene (BHT)

Function
Preservative-Antioxidant

BHA

BHT

Description

Like a team of dedicated bodyguards, these two compounds give themselves up in order to protect their bosses: processed food products. It's not too dramatic: Both are antioxidants, so their nature is to willingly react with oxygen. Almost any food with fat or oil in it will have antioxidant additives in the mix. The net effect of BHA and BHT's proclivity is that they suck up enough of the available oxygen in a given food container, and fats and oils in the food resist spoiling longer. This in turn protects the food product's odor, color, and flavor. They're so good at this that they're also employed to protect cosmetics (especially lipstick and eye shadow), pharmaceuticals (pills), and other products that are loaded with fats and oils. They even protect rubber. Good thing, because oxidized rubber is no fun.

As with so many chemicals, some of their advantages are also disadvantages. Both BHA and BHT are considered safe by US and EU government agencies when limited to less than 0.02 percent of the food's oil or fat content, and they both also fight some illnesses (BHT is even part of a research study to fight herpes and AIDS, and in some studies actually decreased the risk of cancer). However, in very large doses, and/or in combination with other chemicals, they could promote some ill health. Some organizations list BHA as a carcinogen, but only based on a variety of situations. BHA has been used in the United States as an additive since about 1947, and BHT since 1954. Oddly, some Web sites say that BHA is banned in the United Kingdom, which is untrue, but its use is narrowly specified and highly regulated. Both are made in industrial specialty organic chemical plants that service not only the food industry but also the broad chemical, plastics, and agriculture industries. The most common raw material for these chemicals is, of course, petroleum.

You'll find BHA and BHT protecting very common foods, including shortening, butter, meat, cereal, chewing gum, snack foods, and dehydrated potatoes. Sometimes they are also included in the paper packaging of the foods they protect, presumably because fats can soak into cardboard and also because they stabilize the wax coatings of packaging (which, like BHA and BHT, are petroleum products). In fact, they are commonly used to stabilize paraffin, essential oils, polyethylene, polyester, and polyurethane. They are at work in insecticides, synthetic lubricants, and even inks and paints. The power of antioxidants is far-reaching.

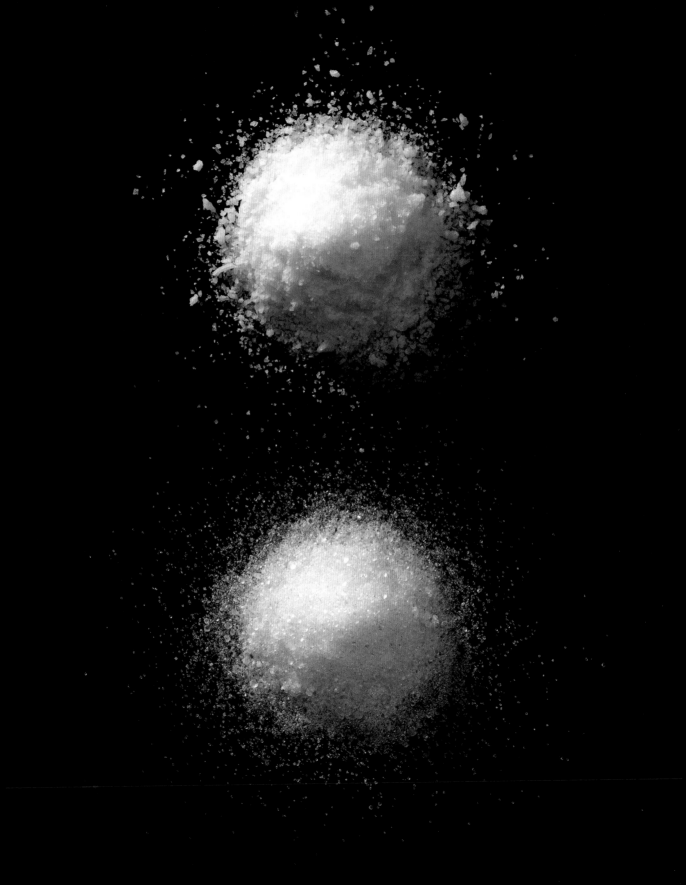

Caffeine

$C_8H_{10}N_4O_2$

CFR number
182.1180

E number
E300

CAS number
58-08-2

Synonyms or siblings
Guaranine, Methyltheobromine, Theine, 1,3,7-trimethyl xanthine

Function
Appeal-Flavor, Stimulant

Description

Caffeine is familiar as a welcome component of our day, the chemical that puts the oomph in our favorite morning or after-dinner beverages, coffee and tea. But vendors of caffeine for processed food describe it less poetically. "Caffeine is a bitter, white crystalline xanthine alkaloid that is a psychoactive stimulant drug," according to one catalog. It is also known as guaranine, methyltheobromine, theine, trimethylxanthine, and mateine, according to others. Not very appetizing names.

Though caffeine is also found in guarana seeds (used widely as a caffeine source in Brazil) and kola nuts (now an uncommon source, from West Africa; genus Cola), most of what you buy these days in a product such as an energy drink is simply the stuff that's removed from coffee in the process of making it decaffeinated. Coffee companies remove the caffeine using a few methods that involve water (most commonly), CO_2, or organic solvents. They dry it, package it, and sell it off to soft drink and other manufacturers.

Caffeine is stimulating—in this case that means giving a wake-up call to the central nervous system, heart, and respiratory system. It can alter blood sugar release and calcium excretion, and it often acts as a diuretic. It affects people differently, especially given a person's age. Many consider it the leading psychoactive drug in existence. Its psychoactive quality even plays an important role in nature. The caffeine available from the plants that grow it (coffee, tea, yerba maté, guarana, cacao) is so stimulating of certain insects' nervous systems that it works as a pesticide, paralyzing and killing insects on the spot.

The discovery of caffeine as the chemical behind coffee's, shall we say, inspirational effect on so many writers and talkers was fittingly motivated by a renowned coffee-loving writer, Johann Wolfgang von Goethe. The great poet was quite interested in science and, in 1819, visited a lab in Germany where an assistant named Friedlieb Ferdinand Runge impressed Goethe with his discoveries of various chemical compounds and processes, including one that made his cat's eyes dialate. Inspired, Goethe gave Runge some coffee beans and challenged the young man to find out what chemical was behind coffee's special effects. Within months Runge managed to isolate caffeine as the active ingredient. A year later three French chemists also isolated caffeine, and, being French, named it after the most familiar source, the French word for coffee: café.

Although coffee has been known since the ninth century (legends trace it to being discovered in Ethiopia), caffeine extract has been manufactured only since 1821, when some French pharmacists took Runge's discovery a step further and figured out how to extract it on an industrial scale. Happily, they chose not to patent the process; one of the scientists, Joseph Bienaimé Caventou, got a crater on the moon named after him for that kind of open-source thinking.

Calcium sulfate

$CaSO_4$

CFR number
184.1230

E number
E516

CAS number
7778-18-9

Synonyms or siblings
Calcium sulphate, Anhydrite, Drierite, Anhydrous calcium sulfate

Function
Appeal-Color
Process and Prep-Improve Powder Flow, Bleaching Agent, pH Control Agent, Stabilizer

Description

Here's yet another rock that we eat. In spite of its chemical name, and the fact that its sister material, plaster of paris, is used to make casts and walls, calcium sulfate shares the distinction (along with eggs, salt, and water) of being the least processed of popular food ingredients: We do hardly anything to it before eating it. Like salt, it comes from an ancient marine deposit near the Earth's surface and it is essential to our health.

Although calcium sulfate is one of the most common minerals in the world, nowhere else is it found—and used—in such quantities as in the United States. Calcium sulfate is processed into various versions, with slightly different amounts of water contained in each molecule (less in plaster, more in the kind used in food). Quite often, despite not always being accurate, all forms of calcium sulfate are referred to as gypsum. The purest gypsum (the only gypsum approved for use in food) is found in the rolling Gypsum Hills of Southard, Oklahoma, where the land sparkles with glasslike crystals. It seems that even the Earth there is calcium-enriched. One quarry yields just about all our food-grade calcium sulfate. It is so pure (98 percent to 99 percent) that the stuff is just scraped off the ground by giant bulldozers, ground into a powder on the spot, and simply dried. It remains a natural ingredient.

According to legend, Benjamin Franklin is responsible for the widespread use of plaster of paris as a soil amendment in the United States (it promotes aeration in clay soils). In the 1880s, a man named Augustine Sackett figured out how to dry a layer of plaster between two pieces of thick paper, inventing gypsum wallboard. It really took off on a national scale 50 years later, when his Sheetrock was featured at the 1933–34 World's Fair in Chicago. Now it seems that every wall is made of it.

Calcium sulfate was officially approved for food use in the United States in January of 1980, but we're way behind the rest of the world in that way. Gypsum has been used in

The purest gypsum (the only gypsum approved for use in food) is found in the rolling Gypsum Hills of Southard, Oklahoma, where the land sparkles with glasslike crystals.

food for more than 2,000 years. In China, it was used during the Han dynasty for coagulating soy milk to make tofu (and still is); some say it may have been used to stiffen bread dough in King David's Jerusalem. Flour that is low in calcium can produce soft, sticky dough, which would frustrate any baker and be a nightmare for the major bakeries with all their pumps and machines; that's why calcium sulfate is found in most types of regular bread. Added in minute percentages—never more than 1.3 percent by law—calcium sulfate is loosely categorized as a dough conditioner for baked goods, which is its most common food function.

Some suppliers have identified more than 100 uses for calcium sulfate as a direct food additive, including the very common one of acting as a filler, keeping custom-mixed-baking powder ingredients mixed evenly and preventing caking. It can feed yeast, balance acidity, firm up ice cream, bind with pectin to keep canned fruits and vegetables juicy, whiten cake icings, and work as an abrasive and an alkali in toothpaste. Beer brewers use it not only to raise the pH of beers that are too acidic but also to control proteins and starches to make the beer smoother and more stable on the shelf.

Calcium sulfate often figures in a recipe merely to add calcium as a nutrient. When it's as cheap as common salt, why not?

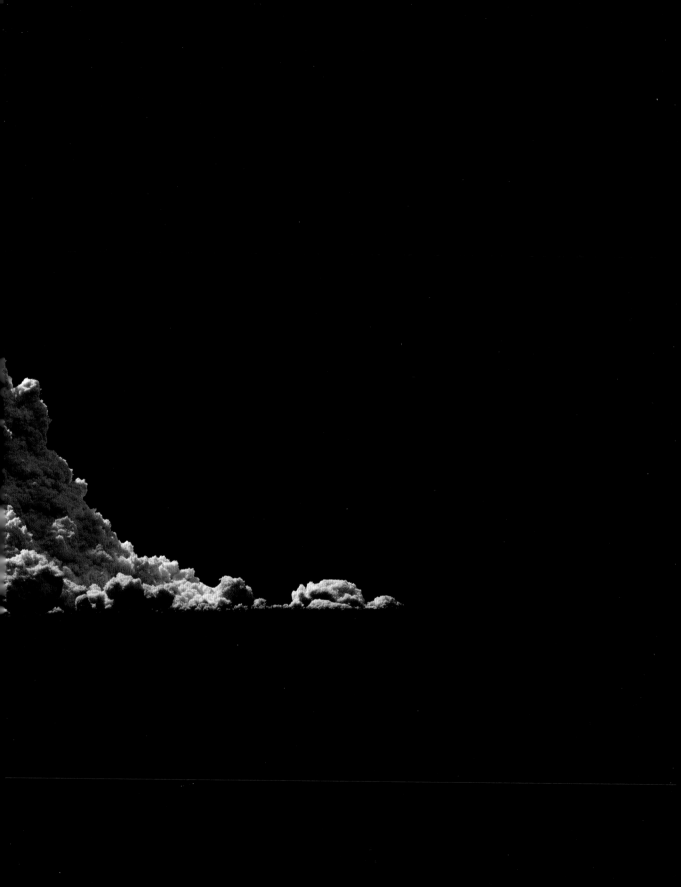

Caramel color

CFR number
73.8500

E number
E150a-d

CAS number
8028-89-5

Synonyms or siblings
Caramel coloring, Plain caramel, Sulfite caramel, Ammonia caramel, Sulfite ammonia caramel

Function
Appeal-Color

Description

One of the few food ingredients anyone can make at home is also the most widely consumed food coloring in the world—it makes up 80 percent of all colorants, with more than 200,000 tons produced per year and is one of the rare ingredients to be manufactured on five continents. Caramel coloring darkens coffee, soups, gravies, baked goods, soy sauce, ice cream, candy, and even some meat. It is usually listed by name but also is one of those unidentified "natural and artificial colors" because though it's processed, it contains only cooked sugar. As most cooks know, many foods brown when cooked; that's usually just the food's natural sugars geting caramelized.

Manufacturers make four different kinds of caramel color, hence the four different "E" numbers (150a, b, c, and d). They each behave differently in various foods. Each kind is made by heating sugar in the presence of ammonia and sulfur compounds, acids, salts, or bases. The four types vary in atomic charge, viscosity, pH, and so forth. That means that some are better in solid foods, some are more suited for different beverages, some create a haze in liquids, and some react badly with acidic foods. It should be noted that the same colorant that shades commercial eggnog also makes precooked microwaved poultry look oven-roasted. Caramel coloring is so concentrated that it is used in minute amounts that have no effect on the calorie or nutrition profiles of food products.

Most of the world's caramel coloring output is used to color soft drinks, especially colas. In 2012, some caramel coloring manufacturers altered the process for making the specialized version used in soft drinks in response to some consumer groups' demands. This was despite the FDA's assertion that a toxic dose of the colorant would require drinking more than 1,000 cans per day. Industrial coloring is made by heating some form of corn sugar [p. 52], such as dextrose [p. 62], invert sugar, or molasses, under pressure, allowing it to first dehydrate and then condense into more complex stuff. The lighter (and sweeter-tasting) colors are cooked less, the darker (and more bitter) ones are cooked more. The various reactants mentioned above are introduced at this point.

Brewers and distillers were the first food product companies to use caramel coloring, putting it in dark beers like porters and stouts, and eventually even brandy. It was not until the early 1900s that industrial production of caramel coloring began in earnest, triggered by the development of an acid-stable version for use in cola drinks. Reactions with certain foods are still somewhat unpredictable, so food product manufacturers have to test each batch to see how the coloring is going to work. Whiskey is challenging and vinegar is apparently especially hard to color. And you thought you were seeing only the result of barrel-aging.

Carmine

$C_{22}H_{20}O_{13}$

CFR number
73.1000

E number
E120

CAS number
1343-78-8

Synonyms or siblings
Cochineal, Cochineal extract, Carminic acid, Crimson lake, C.I. natural red 4

Function
Appeal-Color

Description

One would not assume that the Italian aperitif Campari and boysenberry yogurt share the same coloring, nor would one assume that such a coloring would come from a scale insect fond of a particular cactus plant. Regardless, all of these things are true.

The fascinating, rich magenta carmine, also called cochineal (of which carmine is the purified version), is one of about 25 so-called natural colors used in foods today. Extracted from the dried body and eggs of the female cochineal insect, which accumulates on the paddles of the prickly pear cactus (genus Opuntia) native to Mexico and South America, one pound of this colorant requires about 70,000 insects. They are simply brushed off the cactus, dried, and mixed with aluminum or calcium salts.

Carmine's use dates back to the days of the Aztecs; some traditional Oaxacan artisans still use it to dye their handmade textiles. When Oaxaca, Mexico, was the center of Mexico's carmine monopoly in the 1700s and early 1800s, carmine exports rivaled silver in value due to the Spanish royalty's love of the deep red color for their garments. These days its value might be less, but carmine is still popular for use in foods and cosmetics. Large carmine plantations are now found in Peru (the world's primary supplier), Guatemala, and the Canary Islands. The output of the Canary Islands is used almost exclusively to color Campari.

Carmine is listed under a variety of names, including cochineal extract, crimson lake ("lakes" are powdered versions of colorants), natural red 4, and C.I. 75470. Colors derived from natural sources are often processed in the same plants as purely chemical ones, and because they have been so manufactured (or synthesized, in the case of beta-carotene), they are simply no longer considered natural. A label describing these colors can say, "color added" or "artificial color added" or actually name the color, but it can't say "natural color." The FDA still classifies colorants as artificial unless they are coloring the very food they come from, e.g., strawberry juice added to strawberry ice cream. But carmine, the ingredient that involves cacti and beetles, that colors an Italian aperitif and yogurt, is an exception: all natural (though not vegetarian). Who knew?

Carrageenan

CFR number
172.6200

E number
E407

CAS number
9000-07-1

Synonyms or siblings
Carrageen, Chondrus extract, Irish moss gelose, Sodium carrageenan, Eucheuma

Function
Process and Prep- Emulsifier, Stabilizer, Thickening Agent

$(C_{12}H_{17}O_{13}S)n$
Kappa carrageenan

Description

If you eat lots of ice cream, then you are eating lots of seaweed. That may seem confusing; we're familiar with seaweed salad in Japanese restaurants, but as a common, everyday food additive its reputation remains nebulous. In fact, the confusion gets worse as you look into the name itself. Carrageenan is an Irish word introduced around 1829, derived from "carraigin," which is variously defined as "moss of the rock" or "little rock." "Carrigan" or "Carrageen" is a place name found throughout (presumably coastal) Ireland, and it could go back to Carrigan Head in County Donegal in northwestern Ireland, but that spot appears to be wild, with no local seaweed industry (Ireland does export seaweed from elsewhere). Carrageenan, specifically the extract of Chondrus crispus, is best known as Irish moss. Now, almost all of it comes from seaweed farms in the Philippines that grow it on lines strung between floats, with lesser amounts coming from China, Japan, and other Southeast Asian nations. The industry is concentrated in Southeast Asia, although seaweed is naturally plentiful around the world. And despite all the Irish connections, not one major seaweed processor now has an Irish address. The industry has migrated over time.

As with alginate [p. 6], years ago the Irish harvested carrageenan mostly as sources of iodine and potash. Back in the early 1800s, they were the first Europeans documented to popularize using it as food—making a sweet milk-based pudding and a cure for respiratory ailments. Immigrants probably brought this know-how with them to North America during the mass exodus from Ireland in the mid-1800s. A strong, sustainable wild seaweed-harvesting and -processing industry developed in Nova Scotia, Canada, that persists on a smaller scale to this day. When World War II interrupted the supply of agar [p. 4]—which had been mostly a Japanese export—the coastal US and Canadian firms expanded, unknowingly laying the groundwork to supply the post–World War II food-additive demand. In the 1970s, these North American firms could not keep up with demand, inspiring industry development in Southeast Asia. In 2009, the Philippines alone exported approximately $200 million worth of carrageenan, creating an industry that is credited with employing many former armed insurgent soldiers along the otherwise

If you eat lots of ice cream, then you are eating lots of seaweed. That may seem confusing; we're familiar with seaweed salad in Japanese restaurants, but as a common, everyday food additive its reputation remains nebulous.

underdeveloped coast. However, since then, China has taken aggressive steps to supply seaweed to processors, creating yet another move in the seaweed cycle that started in Ireland, went to Canada, moved to the Philippines, and now on to China.

You can easily extract carrageenan from red seaweed (the industry uses several varieties) at home by simply boiling it while wrapped in cheesecloth. The carrageenan dissolves into the water and forms a gel when it is cooled. The industrial process is more complicated. The boiling process includes washing and cooking in an alkaline water and sometimes precipitation with isopropyl alcohol or a salt. "Natural grade" or "semirefined" carrageenan skips these steps and instead the impurities are extracted. In both types the gel is dried, concentrated, and powdered. The result, in technical terms, is polysaccharide sulfate ester.

Ice cream is perhaps the most recognizable product made with carrageenan, but any cream or milk drink, especially chocolate milk, needs it to keep the various fats and proteins bound up. Soy milk relies on it for both a smooth mouthfeel and to keep various layers bound together. It's also used in salad dressings, cheese spreads, frozen fruit syrup, and pressurized whipped cream. Beer brewers rely on it to bond with excess proteins left over from fermentation (to elminate cloudiness, or "chill haze"), which can then be filtered out.

Carrageenan is not only a versatile resource in foods and food processing, it is an excellent fat substitute and emulsifier in nonfood products, such as cosmetics, pet food, toothpaste, air fresheners, and shoe polish. It's the ingredient that creates the marbling effect of ink on paper or fabric and makes pills and tablets easier to swallow. Long used in organic fertilizers (for its mineral content), its potential as a biofuel source is currently being explored. It has some antimicrobial and antiviral properties, making it a candidate for HIV prevention and treatment drugs. Finally, carrageenan is an extraordinarily good personal lubricant for sensitive body areas during intercourse. All this and more from humble seaweed flour.

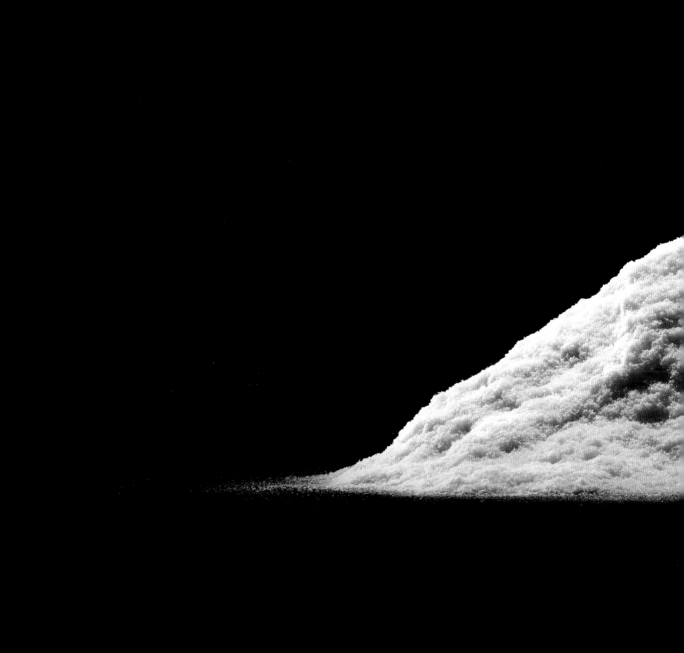

Caseinate (calcium and sodium)

CFR number
182.1748
(Sodium Caseinate)

E number
E469 (Sodium Caseinate)

CAS number
9005-46-3
(Calcium Caseinate)
9005-46-3
(Sodium Caseinate)

Synonyms or siblings
Casein, Milk protein,
Casein sodium

Function
Process and Prep-
Emulsifier, Stabilizer,
Thickening Agent

Description

The very best casein, the protein part of milk that usually gets made into cheese and is beloved of bodybuilders, comes from the Waikato region of New Zealand. In fact, New Zealand is the world's largest casein exporter, half of which goes to the United States. New Zealand is the world's largest casein exporter, half of which goes to the United States. No American dairy can afford to process casein into caseinates, the least profitable part of milk, preferring instead to transform milk into cheese and its popular, pricier by-product, sweet dairy whey [p. 182]. Much of the lower-grade sodium and calcium caseinate comes from Belarus, Russia, and Poland.

In New Zealand and Ireland (by far our two biggest suppliers), milk is sometimes converted to dried caseinate right at the dairy. A bit of hydrochloric acid (instead of an enzyme like rennet) separates milk into curds and whey, only this time the curds get made into cakes, bread, and energy drinks instead of cheese. In other countries, the milk is dried first and has to be rehydrated back into liquid once it's imported.

The drying process is actually pretty simple. Washing curds with a dilute alkali solution—made with either a sodium solution (this would-be sodium hydroxide, or lye) for sodium caseinate, or a lime solution (calcium hydroxide, or lime) for calcium caseinate—neutralizes the acid milk product, turning it into a more useful salt. After the wash, both kinds of caseinates are spray-dried: They're atomized under high pressure into a stainless steel spray-drying chamber that can be as much as five stories high and 375 degrees Fahrenheit hot. There, the liquid caseinate hits the heat and instantly drops to the bottom of the chamber as a fine, pale yellow powder, ready for packaging and shipping to bakeries. The big plants in New Zealand can top an astounding 30,000 pounds per hour doing this.

In the late 1800s, casein was used to make "milk paint," a product house restorers today still curse as they struggle to remove it; casein was also made into glue (still is, in fact), which might explain the paint's tenacity. It helped form the first plastics in the late 1800s and early 1900s, when it was mixed with formaldehyde and hardened into pens, buckles, and knife handles; casein buttons were the norm at the turn of the 20th century, but not for long.

By the 1950s, casein had been treated as a waste product for some time; it was part of the skim milk left over from

The very best casein, the protein part of milk that usually gets made into cheese and is beloved of bodybuilders, comes from the Waikato region of New Zealand. In fact, New Zealand is the world's largest casein exporter, half of which goes to the United States.

butter-making and in the excess curds from cheese-making.

Much as with whey [p. 182], it was often used for pig feed or simply tossed out. It was around that time when the big food companies were on the prowl for food additives that would increase shelf life and extend the usefulness of, or replace, traditional ingredients such as eggs and milk. As a component in dried milk—which had been on the scene for years—casein was an obvious candidate to explore. With 80 percent of the protein in milk found in the casein, it turned out to be a nutritious option, too.

Low in calories and high in nutrients, caseinates function well as fat replacers. Sometimes calcium caseinate is preferred over sodium caseinate to reduce the sodium content in a food product; it also tends to be chosen where nutrition is a primary concern, while sodium caseinate might be chosen where high emulsification and rapid dispersion are more important. Otherwise the two are quite similar.

Caseinates play an important role in many foods. They help make dough uniform in mass-produced doughnuts, muffins, and waffles, and block the excess absorption of fat in fried foods. They emulsify and stabilize packaged milk shake/milk drink bases, carry flavors and minimize shrinkage in ice cream and other frozen desserts. They also bind and emulsify fat in processed meats (sausage and luncheon meat), and clarify wine (fine particles coagulate with the protein in casein and are easily filtered out). Caseinates help extend eggs (whole, white, or yolk), they support meringue and other foods made of egg whites, and they form the films on edible glazes. They also add some authenticity to analog (fake) cheese and synthetic milk additives (they are rather important in helping coffee creamers disperse quickly). They contribute to stickiness of edible adhesives, like those found on envelopes.

The most fascinating industrial role that casein plays is in making concrete. Part of a new system of products that reduce the amount of water needed so much that the curing time is cut by half, casein is saving oodles of money for builders. But it kind of makes sense, using a glue and nutritious additive to make another material sticky and strong.

Cellulose gum

$C_{14}H_{18}Na_2O_{12}$

CFR number
182.1745

E number
E466

CAS number
99331-82-5

Synonyms or siblings
Sodium carboxymethyl-cellulose, Carboxymethyl cellulose sodium, Modified cellulose

Function
Process and Prep- Emulsifier, Stabilizer, Thickening Agent

Description

Cellulose gum is one of the most abundant, renewable, natural resources in the world. Why? Because all green plants synthesize it to make their cell walls stiff and strong. Processing turns this plant matter into a soft goo that's blended into shampoo, toothpaste, salad dressings, rocket fuel, and "crème" filling.

Most cellulose gum is made from cottonseed fuzz, which is almost 100 percent pure cellulose. Trees grown for paper pulp are the common alternative source. In both cases, the raw material is cooked in a digester vat along with acids and bleaches to break down the lignin that holds the fibers together.

The wood pulp is processed just as it would be for paper, up to a point. Then, the cellulose destined for food use is processed further with bleaches and petrochemicals, completely rinsed, and turned into sodium carboxymethylcellulose, also known as CMC, NaCMC, carboxymethyl ether cellulose sodium salt, carmellose sodium, cellulose carboxymethyl ether, cellulose glycolic acid sodium salt, or sodium carboxymethyl cellulose, among other appellations that you might not find on a menu or in a recipe, but that you'll definitely see on an ingredient label.

Ground into a tasteless and odorless powder, cellulose gum finds its way into countless food products primarily because it can absorb an astounding fifteen to twenty times its own weight in water. This is why cellulose gum is used to thicken liquids, stabilize blends that would otherwise separate, and bulk things up a bit. It's a major player in bottled salad dressings and canned soups. Quite possibly, it could be a secret ingredient in the famous Twinkie creamy filling, but no one's talking, that's for sure.

A lot of food scientists love cellulose gum because it can actually replace fat in a recipe, as far as mouthfeel and appearance go, but it doesn't add a single calorie because it's indigestible. It keeps things smooth, thick, and even shiny, much like butter or oil.

Although it seems like an obvious choice for a food product (because it's a plant), cellulose gum was first developed by German chemists making cellulose acetate for photographic film. They went on to try to make artificial silk, which led to the invention of rayon in 1918 (it was not produced on a major scale until 1934). Over in the United States, Dow Chemical started using cellulose to make cellophane, a precursor to plastic, in 1927. Cellulose gum moved from an afterthought of a plastic product to a powerhouse food product only in the 1960s.

Chlorophyll

CFR number
73.125
(Sodium copper chlorophyllin)

E number
E141(i)
(Copper complexes of chlorophylls)
E141(ii)
(Copper complexes of chlorophyllins)

CAS number
11006-34-1
(Sodium copper chlorophyllin)
479-61-8 (Chlorophyll a)

Synonyms or siblings
α-Chlorophyll, Sodium copper chlorophyllin, Natural green 3, Copper phaeophytin

Function
Appeal-Color

$C_{55}H_{72}MgN_4O_5$
Chlorophyll a

Description

Think back to your high school biology class (if you remember it), and you might recall learning that chlorophyll is a pigment essential to photosynthesis—the process by which all plants convert light into energy, the original solar power. Think back to your childhood (or any sports-filled weekend) and picture grass stains on your knees. That's chlorophyll, too. Once extracted from grass, nettles, or spinach, it is transformed into a waxy blue-black microcrystalline pigment suitable for food use. Spinach, by the way, has more chlorophyll than any other common vegetable.

For a natural color, the extraction process is alarmingly unnatural, employing solvents such as acetone, ethanol, and light petroleum (the same process is used for most chlorophyll intended as a nutritional supplement). In 1817, two French pharmacists, Joseph Bienaimé Caventou and Pierre Joseph Pelletier, were the first to extract chlorophyll using solvents. These same pharmacists managed to extract caffeine [p. 26] from coffee and tea in 1821. Because of chlorophyll's fairly unstable nature, it wasn't used as a food ingredient until the late 1990s.

Even though chlorophyll is found basically everywhere there is life, it is strange that no animal uses or produces it. None, that is, except for a certain kind of green sea slug that can digest and use chlorophyll for its own photosynthesis, a process fetchingly and yet alarmingly named kleptoplasty.

Because it is unstable (it oxidizes rapidly), some chlorophyll is hydrolyzed and treated with copper to become a green-black powdered salt called chlorophyllin or Natural green #3 and given a separate E-number, E141(i) for water-soluble and E141(ii) for oil-soluble mixes. Either way, it is still not particularly stable, but as a dye it is helpful in certain chewing gums, ice cream, pasta, soups, and sweets. Perhaps its most distinctive use is as the green fairy colorant in absinthe. Chlorophyll is often used to dye waxes and oils as well as medicines, cosmetics, and soaps. It is also marketed, when highly diluted, as a nutritional supplement with claims that it might prevent cancer, detoxify the liver, and eliminate bad breath, among other improbable acts.

As with so many natural colors, chlorophyll does in fact have some beneficial side effects—in this case, a particularly nice one: it can fight cancer. But this really only holds true when it is injested as part of a green vegetable. You can definitely say chlorophyll is more than safe. And eat your spinach.

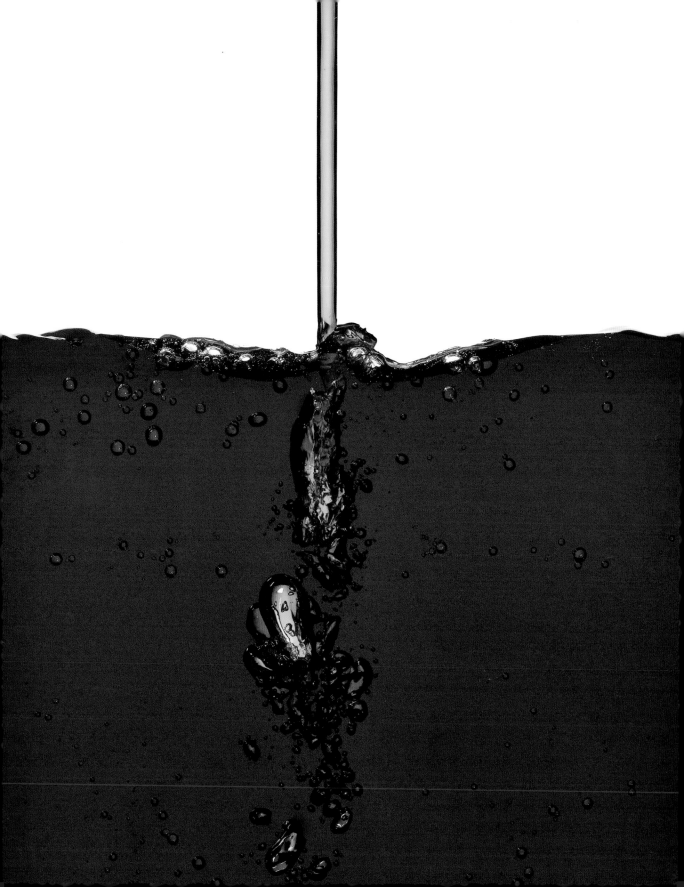

Citric acid

$C_6H_8O_7$

CFR number
184.1033

E number
E330

CAS number
77-92-9

Synonyms or siblings
2-hydroxypropane-1,2,3-tricarboxylic acid, 3-carboxy-3-hydroxypentanedioic acid

Function
Preservative-Antioxidant
Process and Prep-pH Control Agent, Stabilizer

Description

Citric acid is natural to the extent that it's found in all living things (a good definition of "natural"). It is a widely used acidulent and pH control agent in the food product and beverage industry. If you need an acidic zing in your thing, grab the citric acid.

You can easily find this fruit acid in any citrus fruit, but it's also prevalent in strawberries, bananas, and milk. For hundreds of years we simply squeezed lemons and processed it out of the juice, but meeting the demands of the modern food industry calls for specialized companies. Today, we ferment the citric acid from sugar mixed with microbes and turn it into a fine powder in a simple metabolic process called the Krebs cycle. Most citric acid is now made in China at the rate of about 2 million tons per year. The sugar, referred to as a carbohydrate substrate, is most often sugar beet molasses (or corn sugar sources). The fermenting agent is air plus the mold Aspergillis niger, which generates CO_2, mycelium (a sort of fungal root system), and citric acid. The fungus is served to cattle and the citric acid to you and me.

Ever since an 8th century Muslim alchemist discovered it, people have been using lemon and lime juices for their acidic nature. It wasn't until 1784 that citric acid was isolated from lemon juice, by Swedish chemist Carl Wilhelm Scheele, and for more than a century it was an agricultural product largely made in Sicily. That was a problem during World War I, spurring the discovery of the biological process in 1917 that remains in use today.

About half of current production goes into beverages, a fifth into food, and about a third into cosmetics, pharmaceuticals, and industry in a vast array of uses. In food, it provides tartness, reduces sweetness, enhances the effectiveness of preservatives and gel strength, acts as an antioxidant, gives a boost to leavening, and stabilizes color, taste, flavor, and vitamins. It's in foods like jellies, baby food, meat products, fish products, canned food, candies, and dairy products. Citric acid goes into all kinds of beverages, ranging from juices to teas and even wine, but it's most common in soft drinks.

Citric acid plays just as extensive a role in pharmaceuticals, providing effervescence, buffering pH, and even acting as an anticoagulant in chewable and dissolvable tablets, syrups, and suspensions. It serves much the same purpose in personal care products such as shampoos, astringent lotions, toothpaste, and mouth rinses.

Beyond the personal, it is an important part of dishwashing liquids and powders, drain cleaners, fabric softeners, laundry detergents, and surface cleaners. In industry, it is in adhesives, animal feed, concrete and plaster mixes, dyes, papermaking, textile manufacturing, and even both wastewater treatment and water conditioners. With citric acid, you get a whole lot more than just simple lemon juice.

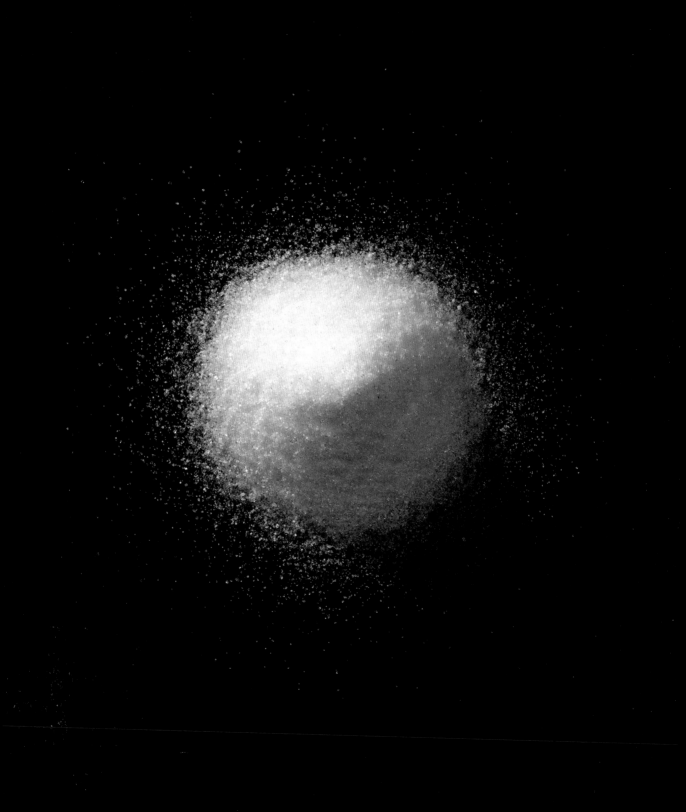

Corn Ingredients

A Mini-Primer on Corn Ingredients

More food additives are made directly from corn than from any other raw material except petroleum; it is the most important crop in the United States. Almost all of it is a starchy, thick-skinned kind called #2 yellow dent corn or field corn. Despite being destined to be used in many food ingredients, it is inedible.

Most corn is processed in one of two ways: dry milling and wet milling. Dry milling has roots in an ancient Native American technique in which kernels are soaked in a hot water and caustic soda (lye) solution to soften them up. The kernels are then gently crushed so the germ can be removed for making corn oil (except for corn destined for real corn chips). The mill crushes the remaining parts of the kernels into descending levels of coarseness: coarse grits (hominy), flaking grits (for cornflakes and other breakfast cereals), brewers' grits (for whiskey, beer, antibiotics, industrial enzymes, and even vitamin C), cornmeal (in various grinds for bread, muffins, polenta, and dusting baked goods), and ultimately, fine flour (the only item included here). Various grades of crushed kernels also go on to become cattle feed, diesel fuel (biodiesel), and explosives.

Wet milling is the first of many complex steps toward making either starches or sweeteners. The wet milling process is based on the acid hydrolysis technique developed in Europe in 1811. Nowadays, manufacturers still use acid but with molecular biology and rooms full of computers to get a variety of different sugars with different characteristics for a variety of products. (Potatoes, rice, and wheat are

also common starch sources that are sometimes blended, which is why food labels often say just "food starch.") We include seven wet-milled corn ingredients here.

The corn is cleaned and steeped in warm water infused with a touch of sulfur dioxide to keep it fresh. The soggy kernels are centrifuged to separate out various components. The remaining starch slurry is pumped into the corn refinery and washed up to fourteen times to remove the protein and produce 99.5 percent pure starch. This starch slurry is the foundation for the two branches of corn processing that eventually yield either starch products or sweeteners.

At a starch plant, the starch slurry is either dried, chemically modified, or roasted into the three most common products (cornstarch, modifed cornstarch, and maltodextrin), but some plants make more than 400 different kinds of starches for industrial and food uses.

At a corn sweetener plant, starch slurry is mixed with hydrochloric acid and/or enzymes such as alpha-amylase along with a dash of lime or caustic soda (for pH control) to convert the slurry into a low-glucose solution. Other enzymes convert it further and the process is repeated under exacting, tightly monitored control until the desired sweetness level is reached, creating the different ingredients pictured here: dextrose, corn syrup, and high-fructose corn syrup.

Corn flour

CFR number
137.2110

CAS number
68525-86-0

Function
Process and Prep-
Thickening Agent, Powder Flow Agent

Description

Corn flour is the odd man out among corn ingredients. Technically speaking, it is neither a thickener nor a sweetener, though it does have some similar attributes ("cornflour" is the British term for what we call cornstarch [p. 56]). Corn flour is dry milled.

Corn flour is also an important addition to the wheat flour used in making baked goods; it blends well with most any powder. It provides a tool for reducing gluten content. And it provides a unique gel-like texture to the interior that holds both the structure up and the moisture in, making simple old corn flour yet another additive that helps prolong shelf life.

Cornstarch

CFR number
172.892

CAS number
9005-25-8

Synonyms or siblings
Cornflour starch, Maize starch, Unmodified corn starch

Function
Process and Prep-
Thickening Agent, Powder Flow Agent

$(C_6H_{10}O_5)n$
Amylopectin

$(C_6H_{10}O_5)n$
Amylose

Description

In the most basic terms, cornstarch is simply dried starch slurry. This is the stuff you buy and use at home. Food starches have been separated from vegetables since ancient Chinese and Egyptian times, and in the United States since 1848. These days, the slurry is dried in a series of presses, centrifuges, and a lukewarm-air spin around a 150-foot-diameter ring dryer; it can't be heated too much or it will cook into a great, useless blob.

Cornstarch is famous for such pedestrian tasks as thickening sauces and keeping dry mixes dry and flowing. It binds moisture in processed meat and, of course, gives great body to soups, sauces, gravies, fruit pie fillings, and puddings.

Despite being a household staple, only about 7 percent of the 750 million pounds of cornstarch made yearly in the United States goes into food, while a full two-thirds is used to make paper and cardboard. About 27 percent is used to make biodegradable packaging "peanuts," to keep textiles and collars stiff, and to keep baby bottoms dry. Cornstarch products can replace petroleum-based, fossil fuel–devouring products in the form of biodegradable plastic, film, fabrics, carpeting, cups, food containers, and even furniture.

Modified cornstarch

CFR number
172.892

E number
Various

CAS number
Various

Synonyms or siblings
Food starch, Acid-treated starch, Bleached starch, Enzyme treated starch, Monostarch phosphate, Hydroxyproply starch, Distarch glycerol, Acetylated oxidized starch, Pregel

Function
Process and Prep- Moisture Control Agent, Thickening Agent

Description

In the 1960s, big food processors inspired the creation of a chemically altered starch, called pregelatinized starch, or "pregel," which thickens without cooking and stays thick. Officially referred to as modified cornstarch, this popular additive is generally made by mixing cornstarch with propylene oxide (a precursor to polyurethanes and propylene glycol [p. 136]), and phosphorus oxychloride (a hazardous material made from phosphorus, oxygen, and chlorine with the rather cute trade nickname of "pockle") in a violent reaction to make hydrochloric and phosphoric acids and, eventually, modified cornstarch. This spruced-up starch may be bleached or treated with sulfuric or hydrochloric acid—which are easily washed out before the starch is dried—to make such things as canning fruit or pie fillings.

Modified cornstarch is easy to swallow and makes items such as nutritional beverages extra-smooth; it improves the freeze-thaw behavior in microwave meals; it mimics, with few or no calories, that smooth mouthfeel and tongue-coating that used to be the exclusive domain of fat and eggs in desserts and sauces. It's all about moisture control, holding water in creamy fillings. And it needs no cooking.

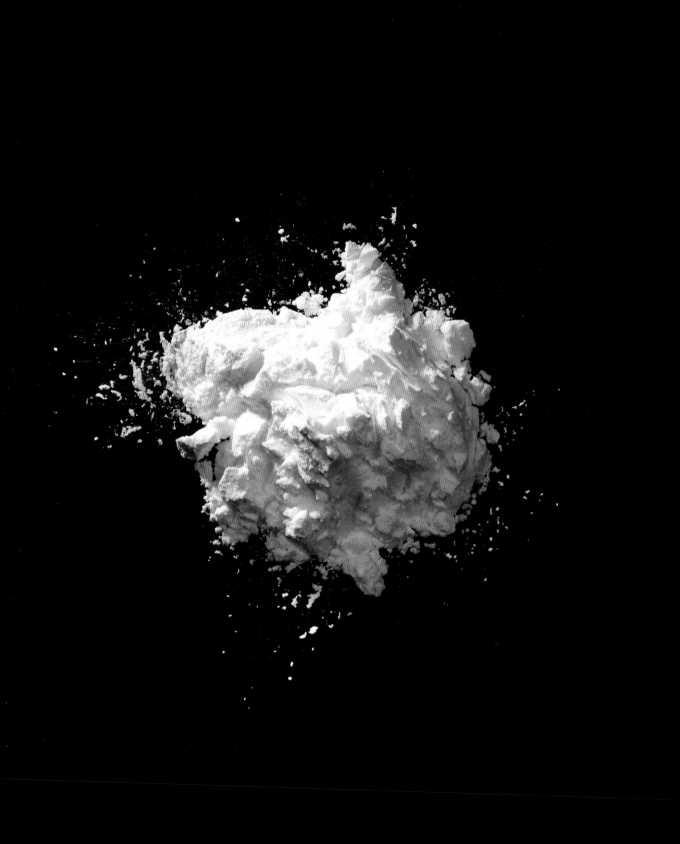

Maltodextrin

CFR number
184.1444

CAS number
9050-36-6

Synonyms or siblings
Dextrin, Dextrin type I, II, III, IV, Dextrin (white), Dextrin 10

Function
*Appeal-*Flavor Enhancer
Process and Prep-
Emulsifier, Stabilizer, Thickening Agent

$C_{12}H_{22}O_{11}$
Maltose

$(C_6H_{10}O_5)n$
Glucose polymer

Description

Maltodextrin, a particularly strong thickener that is usually dried and powdered and functionally similar to cornstarch, is created by processing a starch slurry just a bit less than necessary to make corn syrup (it is made from wheat in Europe, and can be made from rice, potatoes, or barley as well). Almost all of the source's protein is removed in the extensive processing needed to produce it. Like any thickener, maltodextrin can be used as a fat replacer and it is a particularly good one. Some ingredient labels specifically say, when listing maltodextrin, "replaces fat." But it is more than just a thickener.

Because it is a sugar, maltodextrin is often found rounding out the flavor and bulk of sugar substitutes, including calorie-free substitutes, to which it adds just a few calories. It is a favorite texturizer and flavor enhancer in chocolate and candy. It has little to no flavor and only moderate sweetness, and actually has a higher glycemic index than sugar. Brewers use it to improve the mouthfeel of beer. It's especially useful because it's fairly inexpensive.

Surprisingly for something most commonly found in sugar substitutes, nutritional beverages, sauces, and salad dressings, maltodextrin is sold in pure form to bodybuilders seeking a full carb experience.

Dextrose

$C_6H_{12}O_6$

CFR number
186.11

CAS number
50-99-7

Synonyms or siblings
D-glucose, Grape sugar, Corn sugar, Dextrose anhydrous, Dextrose monohydrate

Function
Appeal-Sweetener
Process and Prep-Thickening Agent

Description

A 72-hour enzyme reaction in a room-size vat concentrates a starch slurry into what is called a pure dextrose solution, about 98 percent dextrose. There's a small, three-way fork in the stream at this point, when some of this dextrose is pumped to another part of the plant to be made into high-fructose corn syrup or sorbitol; some is pumped out of the plant to serve as a feedstock for its neighbors making biodegradable plastic cutlery, ethanol (fermented with yeast), amino acids (fermented with bacteria; think monosodium glutamate, or MSG [p. 108], and antibiotics), or citric and lactic acid, among many other things; and some becomes the dextrose found in foodstuffs. Just to keep things confusing, dextrose is often called glucose by industry professionals (almost always so in the United Kingdom), which is actually the form in which dextrose is absorbed into the bloodstream. However, some food labels list dextrose as glucose simply to avoid confusing anyone who uses only that name.

Crystallized dextrose is a bulking agent—an ingredient that adds fewer calories than the ingredient it replaces—that also causes browning through the Maillard reaction (when sugars and proteins react under heat). Dextrose crystals provide a cool feel and a desirable sheen to sweet fillings, ice creams, and frozen desserts. It is completely fermentable (whereas cane sugar is not), so it serves as a successful yeast food in bread and beer. It is used in pharmaceuticals and energy products that demand the swift delivery of energy to the bloodstream.

No matter what form it is in, dextrose/glucose is one of the more efficient humectants, retaining moisture especially well in baked goods. It also rounds out the texture and flavor of cough syrups and lozenges, and helps enhance the flavor of prepared meats. Dextrose adds smoothness, flavor, and shelf life to tobacco, brings glossiness and pliability to shoe leather, stabilizes adhesives, prolongs the setting of concrete, moisturizes air fresheners, and controls the evaporation of perfumes. It helps hand lotion stay moist on your shelf for years—essentially acting as a moisturizer for a moisturizer.

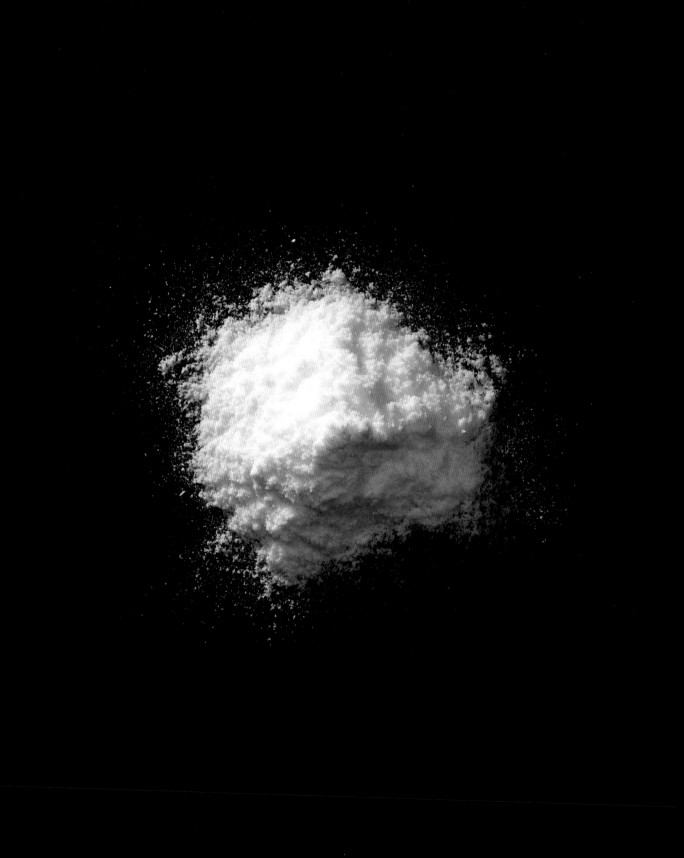

Corn syrup

CFR number
184.1865

CAS number
8029-43-4

Synonyms or siblings
Corn sugar syrup, Glucose syrup, Hydrolyzed starch syrup, Liquid glucose

Function
Appeal-Sweetener
Process and Prep-Thickening Agent

$C_6H_{12}O_6$
Glucose

$(C_6H_{10}O_5)n$
Glucose polymer

Description

Plain old corn syrup is one of the few common commercial processed food additives that is probably on your kitchen pantry shelf. Corn syrup's viscosity, a product of the long molecules that were not "chopped up" by the enzymes or acid, makes it useful as a thickener as well as a sweetener. Because it's only about a third as sweet as table sugar, it's usually paired with other sweeteners in desserts.

Corn syrup also encourages browning (substitute some corn syrup for sugar in your next batch of cookies if you want them to look more toasted). Not only does corn syrup not crystallize like cane sugar, it actually prevents crystallization, which is why it's so popular in candy (it's a common main ingredient), ice cream, and frozen desserts.

Besides baked goods and dessert foods such as fruit fillings, frostings, and cookies, corn syrup can be found in vitamins and as a yeast food for beer fermentation. It even helps suspend the mix of ingredients in salad dressings and hot dogs.

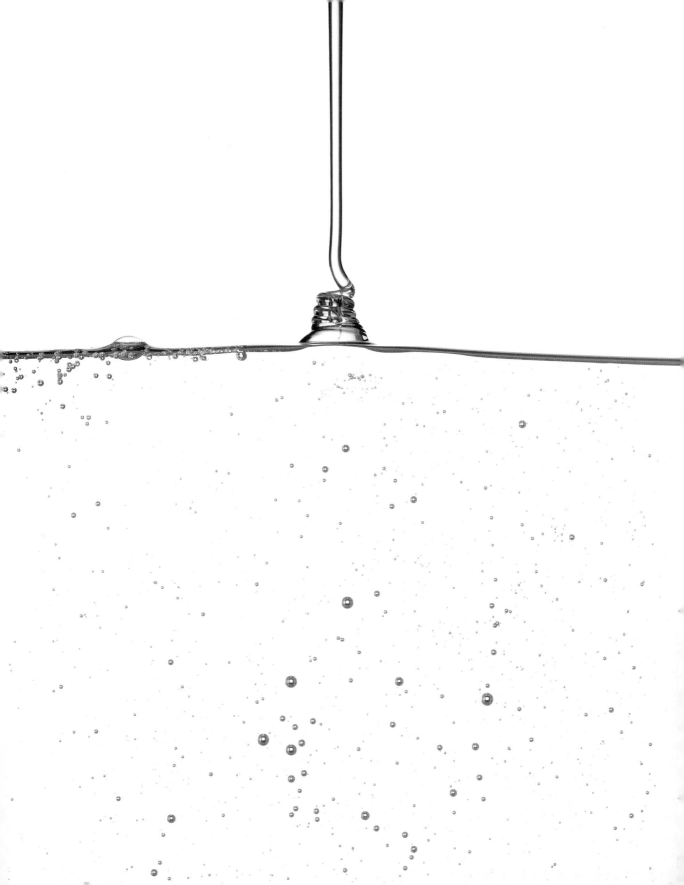

High-fructose corn syrup

CFR number
184.1866

CAS number
8029-43-4

Synonyms or siblings
Glucose-fructose, Isoglucose, Fructose-glucose Syrup, High fructose maize syrup

Function
Appeal-Sweetener

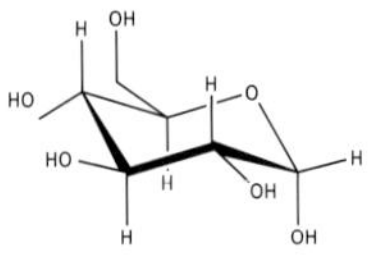

$(C_6H_{12}O_6)n$
Glucose

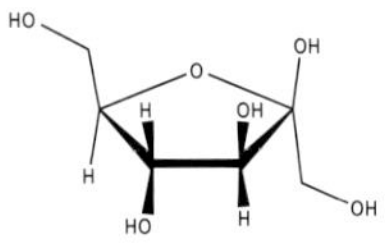

$(C_6H_{12}O_6)n$
Fructose

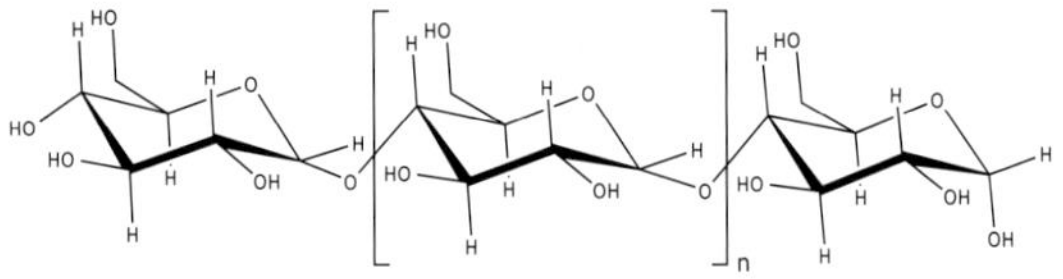

$(C_6H_{10}O_5)n$
Glucose polymer

Description

High-fructose corn syrup is a veritable Zelig of a food ingredient, found in all sorts of products. It's in every kind of fruit, sports, and soft drink, along with wine coolers, yogurts, breads, cereals, cakes, meats, sauces, condiments, and even pet food. This broad range epitomizes popular processed foods. Their makers prefer the stability (it doesn't crystallize), reliable supply, easy mixing, and, of course, the low price of this alternative to sugar in a market where sweet stuff sells best. HFCS and sugar are essentially equally sweet.

High-fructose corn syrup works with its less-sweet but much thicker cousin, plain corn syrup, to perform a variety of important functions: of course it sweetens, but it also gives body, colors via browning, and does a great job of keeping things tasting moist and prolonging shelf life. Industrial bakers rely on it for its ability to soak up and bind available moisture in their fight to resist spoilage. All sugar absorbs (meaning it's more hygroscopic) and binds and holds moisture (as a humectant), but the molecular structure of high-fructose corn syrup allows it to bind water more tightly than other sweeteners do. It is almost twice as absorbant as table sugar. All this ultimately makes factory-baked goods taste moist and fresh, and because it lowers the freezing point of a recipe, also

The syrup undergoes processing (involving fractionation, purification, and recombination) to produce HFCS with different fructose levels, the most common of which are 42 percent (for baked goods) and 55 percent (for soft drinks). Recently, over 46 pounds of HFCS per capita was available in the US food supply.

offers food technologists more options for sweetening frozen food products. No wonder it seems like HFCS is found in almost every processed food product.

HFCS was first made around 1970 but didn't enter mainstream consumer products until about 1980, when Coca-Cola was first swapped for sugar. The manufacturing process involves repeatedly pumping dextrose/glucose through a tower that's several stories high and filled with beads of glucose isomerase enzyme. Each time the dextrose/glucose contacts the enzyme more of it changes to fructose. This "higher" quantity of fructose generates the name "high" fructose corn syrup (plain corn syrup has none or not very much). The higher fructose content allows corn syrup to resemble table sugar, which is 50/50 glucose and fructose.

The syrup undergoes repeated processing (involving fractionation, purification, and recombination) to produce HFCS with different fructose levels, the most common of which are 42 percent (for baked goods) and 55 percent (for soft drinks). Recently, over 46 pounds of HFCS per capita was available in the US food supply.

Sorbitol and mannitol

$C_6H_{14}O_6$ and $C_6H_{14}O_6$

CFR number
184.1835 (Sorbitol)
180.25 (Mannitol)

E number
E421 (Sorbitol)
E420 (Mannitol)

CAS number
69-65-8 (Sorbitol)
50-70-4 (Mannitol)

Synonyms or siblings
D-glucitol, D-sorbitol, Sorbogem, Sorbo (Soribtol)
D-Mannitol, Mannite, Manna sugar (Mannitol)

Function
Appeal-Sweetener
Process and Prep-Powder Flow Agent, Thickening Agent

Sorbitol

Mannitol

Description

First discovered and isolated from the mountain ash berry (genus Sorbus), sorbitol is a sugar alcohol (a polyol) found in a variety of fruits, including apples, pears, peaches, and plums. Usually it is fabricated by hydrogenating a dextrose/glucose solution. We use more than 1 million tons of it a year worldwide.

Sorbitol is a nutritive sweetener—it can be digested and converted back into glucose for energy. However, our digestive system can't quite convert it all, leaving the leftover part free to disrupt our system with gas and diarrhea. Because of this, food formulators use it sparingly. Any food that has 50 grams of sorbitol in a serving must have a laxative warning on its label. On the other hand, that lack of full digestion is good news for people with diabetes.

Sorbitol tastes a little more than half as sweet as sucrose, and it acts pretty much like any other sugar except that it doesn't help with browning. Its main job, whether in food or cosmetics, is to retain moisture, as both a humectant and a thickener (some cigarettes use it as a humectant too). Still, it is often used as a sugar substitute, especially in diet foods (it has a low glycemic index) and cough syrups. Sorbitol is also a common ingredient of mouthwash, toothpaste, and sugar-free chewing gum because it does not cause cavities. Frozen foods use it to protect against

First discovered and isolated from the mountain ash berry (genus Sorbus), sorbitol is a sugar alcohol (a polyol) found in a variety of fruits, including apples, pears, peaches, and plums. Usually it is fabricated by hydrogenating a dextrose/glucose solution. We use more than 1 million tons of it a year worldwide.

crystal formation and general degradation, which is why it is an important ingredient in surimi (imitation crab).

Mannitol functions a lot like sorbitol and is also made by hydrogenating dextrose/glucose. Both are sugar alcohols, share the same molecular formula (though not the molecular structure), and are low-carb foods with low glycemic indexes. Neither digests well, both can give gas or diarrhea, and they are popular sweeteners for products offered to diabetic people.

Mannitol, however, does not create glucose and it is a little less sweet than sorbitol. Instead of occurring naturally in stone fruit, it is found in asparagus, carrots, olives, pineapples, sweet potatoes, and in the seaweed from which it used to be extracted. Consuming more than 20 grams may turn it into a laxative. Above all, mannitol is not a humectant—it does not attract water.

Mannitol is a popular hard-candy coating, non-cavity-forming chewing-gum sweetener, and bulking agent for pharmaceuticals because of its slightly sweet taste and full mouthfeel.

Diacetyl

$C_4H_6O_2$

CFR number
184.1278

CAS number
431-03-8

Synonyms or siblings
2,3-butanedione, Biacetyl,
Dimethyl diketone,
2,3-diketobutane,
Butanedione

Function
Appeal-Flavor

Description

Also known by its not-very-simple names 2,3-Butanedione, Biacetyl, Dimethyl diketone, and 2,3-Diketobutane, diacetyl is one of the more surprising flavorants. It's primarily associated with the smell of butter because it is found naturally there, but it is also quite often included as a part of a well-rounded artificial vanilla flavoring. It does more than just flavor: It can block the CO_2-sensitive neurons in culex mosquitoes, which is bad for the bugs because they basically find you by sensing your CO_2.

In its concentrated form, it totally stinks. Considering that it is the primary element in artificial butter flavor (think movie popcorn and the memorably baffling slug "Golden Flavored Topping"), this would be counterintuitive. But when it is freshly manufactured it must be handled especially carefully due to its intensely disgusting odor. Some producers create dedicated buildings just for diacetyl. You might recognize it as the smell of overfermented beer, spoiled fruit juice, or rancid butter. It is so pungent that it's detectable in concentrations as low as 50 parts per billion, such as the touch of it in a good Chardonnay.

In fact, it is a primary element of the butter smell in nature, and thus could be extracted from butter at great expense; it can also be fermented from yeast. But the same exact molecule is more inexpensively created from natural gas by a few obscure Chinese chemical companies and one well-known German multinational group. All it takes is the biggest chemical plants in the world, a major dose of petroleum, a little hydrochloric acid, plus some very mundane and inexpensive water and air.

Though the recipe is definitely not available for public knowledge, the manufacturers probably start with butane, a natural gas component, and process it with ultrahot steam into a clear, very volatile, flammable liquid called methyl ethyl ketone (MEK). Technically MEK is the main aroma in blue cheese, but don't try telling that to a French *fromager*; it actually smells more like a sweet hospital antiseptic. In the final step, when the MEK is passed over a rare metal catalyst like vanadium oxide and then mixed with a little common hydrochloric acid, it changes into an intensely bright fluorescent yellow, volatile, flammable liquid: diacetyl. Generally, the flavor companies cut this with a number of other chemicals such as butyric acid, delta-dodecalactone, and propylene glycol to make it more manageable and more butterlike. They also might even add vanilla, which is interesting because the resulting blend is often used to round out vanilla flavor. It's all in the family.

Ethylenediaminetetraacetic acid (EDTA)

CFR number
172.120
(Calcium disodium EDTA)

E number
E385
(Calcium disodium EDTA)

CAS number
62-33-9
(Calcium disodium EDTA)

Synonyms or siblings
Disodium EDTA dihydrate, Calcium disodium EDTA, Edetate

Function
Preservative-Antioxidant

$C_{10}H_{12}CaN_2Na_2O_8$
Calcium disodium EDTA

Description

Ethylenediaminetetraacetic acid is a convoluted name almost always abbreviated EDTA. It fights one of the most vexing problems in processed foods: oxidation. While it does act as a preservative by working as an antioxidant, it's also a chelating agent—meaning it chemically sequesters any metal in the food. This is not about errant flakes or machinery parts. Really, trace metals are naturally present in some foods (especially iron and copper) or can get in during processing on machinery (not natural at all). Metal, even in trace amounts, is bad because it oxidizes the food that it contaminates. Oxidation degrades vitamins and leads to rancidity. It's particularly bad news for potatoes and canned crabmeat. EDTA binds with any metals and renders them inoperative.

Its ability to sequester metal ions is essentially why EDTA is one of the most common chemicals used in industrial products; it's especially common in industrial dyes, the paper bleaching process, and laundry detergents (to counteract the presence of metal from hot water heaters). EDTA has a home in hospitals as an antidote to lead and other metal poisoning as well as a possible heart medication. It's widely used in blood banks for preserving blood specimens by working as an anticoagulant.

EDTA is usually used in a salt form, labeled calcium EDTA or sodium EDTA. First isolated in 1935, it is now made in a complex process that includes some random nonfood chemicals such as formaldehyde and sodium cyanide. The resultant salt is harmless, especially because it is used in such small amounts. The same claim cannot be made for its use on a global scale. It is so widely employed that one of its components, when released as EDTA decomposes in the environment and actually becomes an organic pollutant.

In food, its chelating function is critical: It mitigates the formation of benzene, a carcinogen that can form in soft drinks and other foods that contain the common ingredients ascorbic acid [p. 12] (for taste) and sodium benzoate [p. 154] (a preservative). This is important stuff.

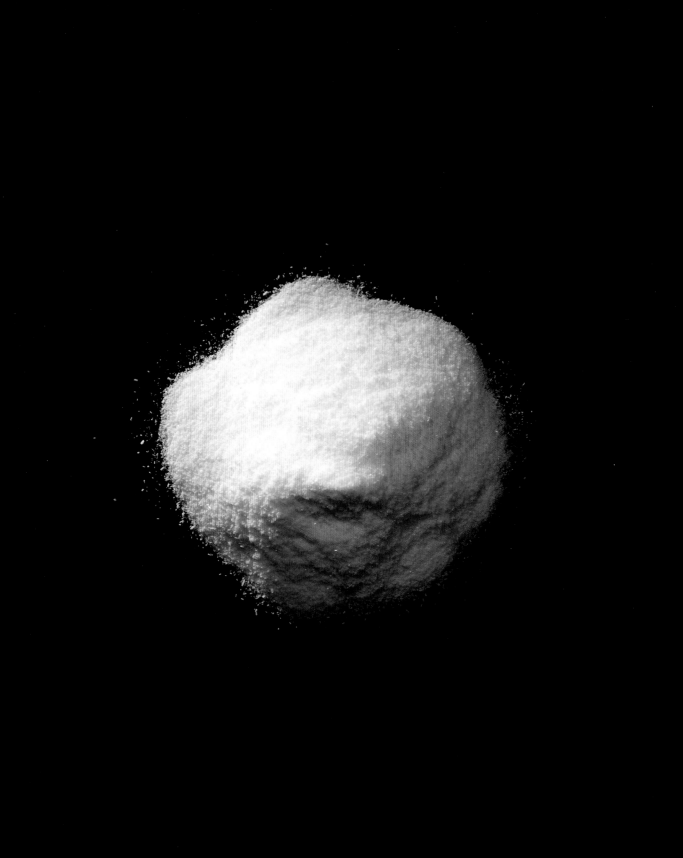

Disodium inosinate

$C_{10}H_{11}N_4Na_2O_8P$

CFR number
172.535

E number
E631

CAS number
4691-65-0

Synonyms or siblings
Disodium 5'-inosinate,
Sodium 5'-inosinate,
Disodium inosine
5'-monophosphate,
Sodium inosinate

Function
Appeal-Flavor Enhancer

Description

Simple and yet complex at the same time, this salt is virtually tasteless on its own. When paired with its relative monosodium glutamate (MSG) [p. 108], however, it multiplies the umami taste effect by a factor of six or even eight. If you see this ingredient on a label, you can be sure that MSG is present in one form or another. They are synergistic flavor enhancers.

Disodium inosinate is simple because, like plain salt, it's a flavor enhancer. This ingredient is a big hitter in food products that need a salty, strong taste. It is likely included in ramen-style instant noodle soups, potato chips, gravies, sauces, dressings, prepared meals, canned vegetables, and fish or seafood.

Disodium inosinate is complex because this chemical, a nucleotide, is a building block in your body, part of the chain of reactions that builds your DNA and RNA. It plays a role in creating your body's chemical energy, and keeps your metabolism going. Too much of it is a problem for anyone suffering from gout; however, since the chemical is used only in minimal amounts in commercial foods, it would be hard to consume dangerous amounts.

While disodium inosinate (inosinate is a salt of inosinic acid) is naturally found in vegetables, mushrooms, and animals (especially sardines), it is normally produced by fermenting corn syrup with some specialized enzymes and a proprietary microorganism that was isolated in the late 1960s. The most common commercial stuff—a colorless or white, stable crystal—comes from Japan (the source of much of the world's packaged ramen, and where umami and MSG [p. 108] were discovered). Whether from sardines or corn, this little salty additive packs a punch.

Since time immemoria
gy and folklore led hu
the mountain ash or E
with its bright orange
It was said to protect a
beings and ward off e
out that the folklore

, Old World mytholo-
mans to believe that
ropean rowan tree,
berries, was magical.
gainst malevolent
l influences. It turns
was right.

Ethyl vanillin

$C_9H_{10}O_3$

CFR number
182.60

CAS number
121-32-4

Synonyms or siblings
Ethylvanillin, Ethyl protal, Bourbonal, Vanilal

Function
Appeal-Flavor

Description

Vanilla is the only tropical orchid in the world that bears fruit. The fruit (vanilla beans) must be hand-pollinated, and for that, workers must wait until the one day in its life that a flower opens. After the seeds are picked, they must be fermented in the ground or dried in the sun over the next three to six months. Then, because so much of the best vanilla comes from Madagascar, Indonesia, or Tahiti, they're usually transported halfway around the world.

Food technologists seeking to understand natural vanilla flavor have broken down its various elements into an almost impossible-to-duplicate 216 compounds. However, we do, in fact, duplicate three of them every day: vanillin (the primary component of natural vanilla flavor), diacetyl [p. 76], and ethyl vanillin. This is the bouquet of natural vanilla.

As the world's most popular flavor and aroma chemical, pure vanilla extract could not meet the world's demand. Hence the need for artificial vanilla. Vanillin was first synthesized in Germany in 1875 from an extract of pinecones. Now it is made from benzene in oil refineries and chemical plants in either China or Baton Rouge, Louisiana. Vanillin is an oxidized alcohol, a (very) distant cousin of formaldehyde. It is not only cheaper to make than it is to harvest the natural stuff, but because it doesn't spoil and is 200 times stronger than natural vanilla, it's cheaper to use as well.

The chemistry begins with toxic, explosive benzene, which is oxidized in a steam cracker, reacted with propylene, and processed into phenol, otherwise known as carbolic acid. For a time, carbolic acid was a sore throat remedy popularized by Sir Joseph Lister, the inventor of surgical antiseptics and the popular mouthwash Listerine. Most phenol is used to make plywood glue, but some is steered into flavoring via another chemical route. The phenol is condensed into catechol, an oily methyl ester, then catalyzed into guaiacol, a light-sensitive alcohol. (Some guaiacol is processed into guaifenesin, the cough medicine, while the rest is processed into vanillin by a high-pressure, high-temperature reaction with glyoxylic acid.)

When it comes to taste, people need variety. Enter ethyl vanillin, also known as 3-ethoxy-4-hydroxybenzaldehyde, with a flavor and odor that's three or more times stronger than regular vanillin. To produce it, catechol undergoes ethylation, creating an ether called guethol that is then condensed with glyoxylic acid and oxidized and processed into the intense flavor ethyl vanillin. It plays a significant part in making perfumes and creating artificial flavors ranging from butterscotch to rum. And it works wonderfully in chocolate.

Ferrous sulfate

FeO_4S

CFR number
184.1315

CAS number
13463-43-9

Synonyms or siblings
Ferrous sulfate heptahydrate, Green vitriol, Iron vitriol, Copperas, Melanterite

Function
Nutrient-Enriching Agent

Description

The touch of iron in enriched flour usually begins not in iron ore mines but in oil wells.

Sulfate, as the name suggests, is derived from sulfur. No longer mined, it's refined out of high-sulfur crude oil in a step that lessens air pollution when the oil is burned. Refineries remove sulfur as a gas, liquefy it into elemental sulfur, and ship it at just under 300 degrees Fahrenheit to sulfuric acid manufacturers, who burn the sulfur to get sulfur dioxide (which is used to process corn into corn syrup [p. 64] and cornstarch [p. 56]). They pass that gas over racks of expensive vanadium catalysts in building-size towers and mix it with water to get sulfuric acid.

With 165 million tons being manufactured each year, sulfuric acid is the most produced chemical in the world. While mostly used industrially, it plays an important role in creating many processed food ingredients such as phosphoric acid [p. 116], lactic acid [p. 98], and artificial vanilla flavoring (ethyl vanillin) [p. 84]. To make ferrous sulfate, the acid is shipped to a steel mill, where iron ore is turned into steel that is squeezed into continuous thin sheets up to 1,400 feet long. A rusty, crusty, oxide scale forms on the surface of the fresh steel and is removed immediately by what is known as the pickling process, which involves running that continuous sheet through sulfuric acid in tubs up to eighty feet long and seven feet wide. The acid is known as the pickle liquor, one liquor not recommended for consumption.

At the end of the day, after thousands of feet of steel roll have been run through the tub and formed into six-foot-wide, seven-foot-diameter rolls, the sulfuric acid left in the tub is saturated with iron and pumped out for separation. Iron sulfate crystals drop to the bottom while the acid is poured off and recycled for further pickling. The crystals are then partially dried into dark, sandy clumps and shipped to ferrous sulfate processors in one-ton supersacs to be dried, purified, and ground into a metallic gray powder ready for the flour mills.

Ferrous sulfate is added to flour to help fight iron-deficiency anemia. However, most ferrous sulfate is used in non-nutritious ways, in products and processes that include fabric dye, ink, water purification, wood preservation, and weed and moss killers.

A common alternative to ferrous sulfate in enrichment blends is reduced iron, which is less expensive but not as strong as ferrous sulfate. However, as it is finely ground (to be more digestible), it becomes more expensive because more of it is needed to become effective and its little specs might darken batter. When flour companies pass their product through strong magnets, looking for errant nuts and bolts, it's possible that the reduced iron dust might pop out, which would be embarrassingly counterproductive, to say the least.

Folic acid

$C_{19}H_{19}N_7O_6$

CFR number
172.345

CAS number
59-30-3

Synonyms or siblings
Folate, Folacin, Pteroyl-L-glutamic acid, Vitamin B9, Vitamin B_c

Function
Nutrient-Enriching Agent

Description

In 1998, folic acid—the synthetic form of folate (vitamin B9)—became the most recent vitamin to be added to the white flour-enrichment mix; it helps prevent spina bifida and other defects of the brain and spinal cords in developing fetuses. As clear a benefit as that is, it took decades to figure out.

In the 1930s, British doctor Lucy Wills found she could cure a certain kind of anemia with the popular food spread Marmite. A decade later, folate was isolated from spinach and named after the Latin word for foliage, *folium*. In addition to spinach, the compound is also found in liver, citrus fruits, nuts, and beans, but much of it is destroyed by cooking. Contrary to what you might expect, it is much better absorbed in the synthetic rather than the natural form, so if you need to supplement, buy a jar of pills rather than a bunch of spinach.

Because it took decades to identify the benefits of folic acid, it wasn't until 1993 that the FDA proposed its addition to the flour-enrichment mix. With overwhelming evidence that folic acid could cut neural tube defects in newborns by as much as 70 percent, millers went ahead and put it in flour well before the mandatory 1998 compliance date—before labels could even be printed—and the FDA was happy to allow it.

Folic acid is mostly processed in China; it's one of the few vitamin manufacturing processes that is relatively clean and simple. Though they keep the actual technique under wraps, the manufacturers reveal that they make B9 with a combination of fermented and petroleum products. It commences with a starch, such as molasses or cornstarch, and involves various amounts of glutamic acid (the basis of MSG), pteroic acid, butryric acid, zinc, and magnesium salts. The result: a fine, light yellow powder that saves lives.

Gelatin is an element in paper coating, printing, and electroplating. Fish skin–sourced gelatin was once prized for etching computer chips.

soups, sauces, flavorings, whipped cream, confections, and dairy products; as a stabilizer in cream cheese, chocolate milk, yogurt, icings, cream fillings, and frozen desserts; and as a kind of glue to bind together layers of confections, decorative sprinkles onto candies, frostings onto baked goods, and seasonings onto meats. And lest we forget, it works as a clarifying agent ("fining") in beer, wine, fruit juice, and vinegar production—it is sprinkled in and binds with whatever is clouding the beverage, settles to the bottom, and then is filtered out. With this breadth of uses, gelatin is too ubiquitous for vegetarians and vegans to avoid completely. There is no true vegetable source of gelatin, though there are plenty of alternative binders, emulsifiers, thickeners, and the like made from corn and seaweed, notably agar [p. 4] and carrageenan [p. 36].

People have been cooking bones for gelatin for a long time. This makes sense since cooking bones is not too far removed from simply cooking meat. Records exist showing gelatin production from as far back as around 1685 in the Netherlands and from about 1700 in England; a patent for gelatin production was first granted there in 1754. It was first produced commercially in the United States in 1808, probably for pharmaceutical use. A US patent for "portable gelatin" was issued in 1845. Since 1932, it's the stuff that hard, two-piece capsules and softgels ("soft elastic gelatin capsules") are made of; in the case of softgels, gelatin is combined with a little propylene glycol [p. 136] and sorbitol [p. 70]. Gelatin is also part of tablets, granules, and other common forms of medicines and cosmetics, most often as some kind of binder. It's used to size artificial yarns, bind abrasives to sandpaper, and make matches. Gelatin is an element in paper coating, printing, and electroplating. Fish skin–sourced gelatin was once prized for etching computer chips. Since 1870, when a British doctor named Richard Leach Maddox discovered gelatin was far superior to the fussy, wet-collodion photographic process, gelatin has been the base of almost all photographic film and paper.

But best of all, it completely melts in your mouth.

Gelatin

E number
E441

CAS number
9000-70-8

Synonyms or siblings
Gelatine, Collagen hydrolysate, Hydrolyzed collagen protein

Function
Process and Prep- Stabilizer, Thickening Agent

Description

Nope, gelatin is not made from horns and hooves! Along with the monosodium glutamate (MSG) "Chinese food syndrome" myth [p. 108], this is one of the most persistent. It is made, however, from cow bones and hides, pigskins and bones, and sometimes even some fish skin. The bones and hides are degreased, dried, and crushed, then soaked in dilute hydrochloric acid, soaked in an alkali (lime, calcium hydroxide, or sodium hydroxide) wash, soaked in acid again, and cooked in hot water. The pigskins are chopped, washed in water, soaked in acid, and washed again. From that point on, both are treated the same, subject to cooking in hot water ("hot water extraction"), producing a resultant slurry that is then filtered, concentrated, sterilized, chilled, extruded, dried, and finally milled into a powder. (Fish skins are mostly just boiled.) That's a long list for what is essentially the extraction of protein from bones and skins in a process that's technically called the partial hydrolysis conversion of collagen. This collagen is the single most common protein in mammals; it's the main organic component of the white connective tissues, skin, and bones of the animals we eat (hooves, horns, and hair are made of keratin, not collagen). We use around 360,000 tons a year of the stuff, a third of which is used in Europe alone.

Gelatin is popular for many reasons, among them the fact that because it consists of 19 amino acids, it is both nutritious and nonallergenic. Besides being a popular dessert with the addition of sugar, flavor, and color (100 million pounds of gelatin desserts are sold each year in the United States), it is used in food products to bind, stabilize, thicken, and clarify things. Gelatin works as a gel former in desserts, lunch meats, consommé, and confections. It works as a whipping agent in marshmallows, nougats, mousses, and whipped cream, and as a protective suspension gel in icings, ice cream, and other frozen desserts, as well as a binding agent in canned meats, cheese, and various dairy products. Gelatin serves as a film on fruit, meat, and deli items; as a thickener in powdered drink mixes, gravies, sauces, soups, puddings, jellies, syrups, and dairy products; and as a processing aid and encapsulator of colors, flavors, oils, and vitamins. It also functions as an emulsifier in cream

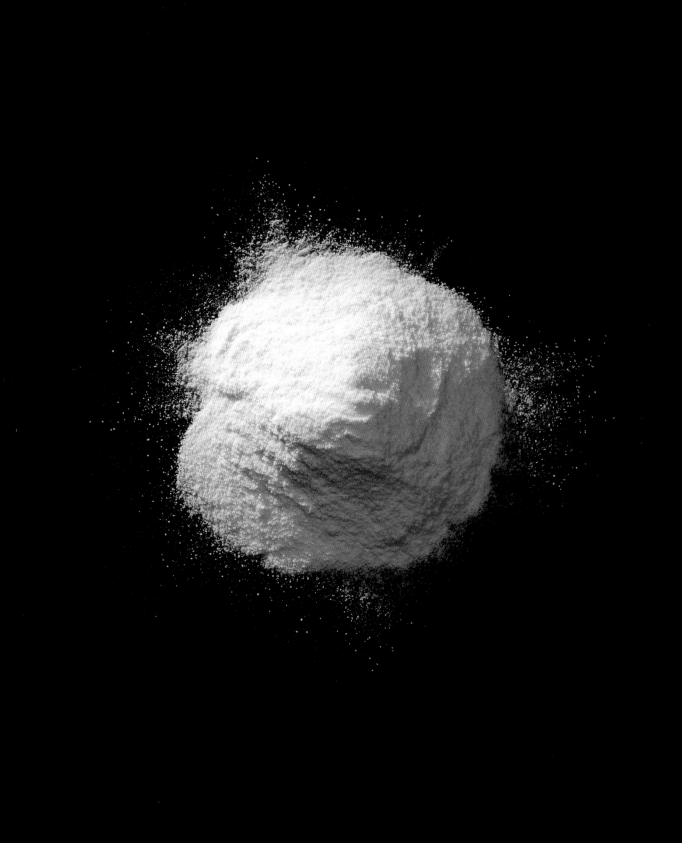

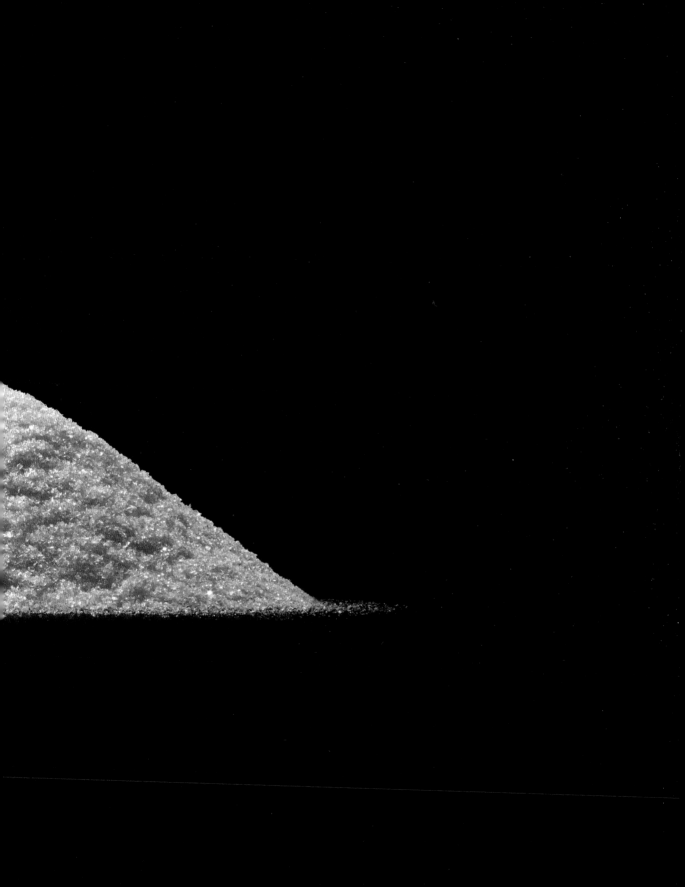

Gum arabic

CFR number
172.780

E number
E414

CAS number
9000-01-5

Synonyms or siblings
Acacia gum, Chaar gund, Char goond, Meska

Function
Process and Prep-
Emulsifier, Stabilizer, Thickening Agent

Description

Many of our most popular pleasure foods are made with a wild gum that is collected by hand from thorny trees mostly in Sudan but also in the Sahel, an area just above the equator stretching from Senegal to Somalia. Gum arabic is sap that gathers in softball-sized clumps called "tears" (trees weep sap). When a small section of bark of the relatively small acacia Senegal tree is removed or slit, workers can often just reach up and grab them; they collect the hardened sap into baskets every ten days during the dry season. Osama bin Laden apparently once owned part of an acacia gum firm in Sudan, but was forced to sell it when Sudan booted him out in 1996.

Processing is fairly straightforward. Bits of bark and sand are removed and the gum is liquefied, centrifuged, filtered, and spray-dried. Much of it shipped out from central African ports to Indian distributors. The fact that so many different cultures have a hand in gathering and processing it may contribute to its many alternative names: gum acacia, chaar gund, char goond, meska harra, and meska.

Gum arabic is one of the most popular food additives around. It's a mixture of saccharides and glycoproteins, used principally as an emulsifier, thickener, or stabilizer, often to simply hold things in suspension. It's also used as a humectant, binder, flavor fixative (in powdered drinks), and a surface-finishing agent. A little bit goes a long way, and you can find it in many beverages (soft drinks, flavored water, beer), snack foods, candies (especially those made with artificial sweeteners), salad dressings, baked goods, and cereals. Usually it must be paired with an antibacterial additive such as sodium benzoate [p. 154].

It's used in a wide range of nonfood items, such as pharmaceuticals, printing inks, paints, textiles, ceramic glazes, drilling muds, and explosives (presumably fireworks), mostly for viscosity control. Gum arabic is used in cosmetics as a binder or viscosity source. It's even part of simple glue found on postage stamps and in incense cones. Traditional medical uses of gum arabic include treating sore throats as well as stomach, intestinal, kidney, and eye problems.

Other gums often substitute for gum arabic, including two from beans: guar gum and locust bean gum. All gums made from seaweed, agar [p. 4], alginate [p. 6], and carrageenan [p. 36] are fine substitutes. Since 1998, xanthan gum [p. 194] also makes the list. Many of these gums are used in gluten-free cooking or to give a "fat feel" in low- or no-fat food products. Their most important job may be the prevention of freezer burn in ice creams.

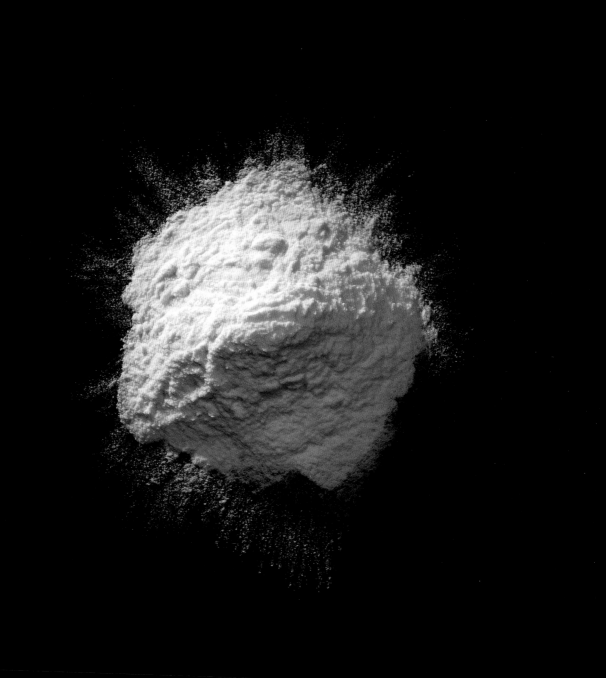

Isoamyl acetate

$C_7H_{14}O_2$

CFR number
172.515

CAS number
123-92-2

Synonyms or siblings
Amyl acetate, Isopentyl acetate, Banana oil, Isoamyl ethanoate, Pear essence

Function
Appeal-Flavor

Description

It may smell like bananas to us, but to bees it smells like pile-it-on party time. This flavor is actually a near duplicate of the pheromone emitted by bees when they find a satisfying target that they just have to share with their buddies. This may partially explain why isoamyl acetate is not used in perfumes.

Often called banana oil or pear essence (it also smells of pears), isoamyl acetate is what's called an ester, a common kind of chemical behind most artificial scents and flavors. Amyl acetate is made in large, high-end hydrocarbon plants, by reacting the industrial workhorse acetic acid with amyl alcohol, using sulfuric acid as a catalyst. (Creating this reaction is also a common high school chemistry lab experiment.) The resulting liquid is fractionally distilled into eight different isomers, one of which is the lovely isoamyl acetate. Although it can mix with water, the ester remains highly flammable and has an irritating vapor in concentrated form. It is naturally formed in ripening fruit with a gentle subtle perfume of what many say invokes the flavor of Juicy Fruit chewing gum rather than actual bananas.

Banana scent or flavor is not just used solo. As with most flavors, blends are what work for our noses. Thus banana oil is part of, in alphabetical order, butter, caramel, cherry, coconut, cola, cream soda, grape, peach, pineapple, raspberry, rum, strawberry, and even vanilla flavorings. It can be found in beverages as well as ice cream, candy, baked goods (it was the original flavor of Twinkies' creamy filling), chewing gum, and gelatin desserts.

Despite its gentle nature, isoamyl acetate plays primarily an industrial role, as a solvent for oil colors, lacquers, and resins. Being both easily smelled and nontoxic, it is also used to test gas masks. It is a great solvent in surface coatings, electronics, leather treatment, and agrochemicals, and is used to make other chemicals. In fact, fragrance uses are essentially an afterthought for this chemical workhorse.

Lactic acid

$C_3H_6O_3$

CFR number
184.1061

E number
E270

CAS number
50-21-5

Synonyms or siblings
L-lactic acid, L-lactate, Milk acid

Function
Appeal-Flavor Enhancer
Preservative-Antimicrobial, Shelf Life Extender
Process and Prep-pH Control Agent, Stabilizer

Description

Lactic acid is natural; it's even produced in our bodies by bursts of muscular activity. It makes your muscles feel sore and tired and gives you cramps in your side when you're running. It also occurs naturally as a result of sugar fermentation in foods ranging from sauerkraut to meat. Lactic acid is responsible for the sour taste in spoiled milk and the sour taste in cheese (Cheetos, too). Unsurprisingly, its name comes from the Latin word *lack*, meaning milk. Carl Wilhelm Scheele, the great Swedish chemist, first isolated lactic acid from sour milk in 1780, and this former waste product has been used to make food ever since. Now, however, the lactic acid made for food use is no longer made from milk—it's made from dextrose [p. 62], a form of corn syrup [p. 64].

While a benign (and secret) strain of bacteria ferments the dextrose in a large tower, the fermentation level of the syrup is controlled by the addition of lime. The process is not all that different from beer-making. The result, calcium lactate, creates a by-product of spent bacteria rich in phoshorus and nitrogen—a great fertilizer popular with local farmers. The next step, called a swap reaction, reacts sulfuric acid with the calcium to produce another popular by-product for the farmers, gypsum (calcium sulfate; useful as a soil amendment), leaving dilute lactic acid. It has to be concentrated and purified, and comes out smelling a bit sweet.

Lactic acid, like salt, brings out savory flavors in beverages and foods and has long been used as a preservative in processed meat and poultry. While helping to enhance flavor, it also stabilizes and preserves salad dressings; it also extends shelf life and helps stabilize the color of pickles, olives, and other brined vegetables. On the sweet side, it is a staple of hard candy and fruit gum. It is a natural sourdough acid and plays a role in cake recipes, too, either as an acid itself or as part of the versatile emulsifier sodium stearoyl lactylate [p. 160]. Despite this long food pedigree, lactic acid is used in a number of fascinating and unpredictable ways in industry, ranging from tanning leather to making CD-ROMs.

Monosodium glutamate (MSG)

$C_5H_8NO_4Na$

CFR number
182.1

E number
E321

CAS number
142-47-2

Synonyms or siblings
Sodium glutamate, Sodium 2-aminopentanedioate, MSG monohydrate, Ve-tsin

Function
Appeal-
Flavor Enhancer

Description

Monosodium glutamate, or MSG, is one of the most famous processed food ingredients. While its scientific name explains nothing to most of us, we've all heard of it, urban legends surround it, it's ubiquitous, and it's also kind of hard to understand what it actually does. For something that's been around for more than a hundred years (it was discovered in 1908), this is rather surprising.

That urban legend thing is big, but silly. Starting around 1968, there were reports that someone (literally just one person) had had a bad reaction to Chinese restaurant food, with symptoms such as flushing, numbness, and a headache. The reaction to the reports ballooned, and came to be called "Chinese Restaurant Syndrome." Some people suggested that the "symptoms" of this made-up syndrome might be a reaction to MSG, and the rumor took off without any scientific underpinning. In fact, while some people seem to be extra-sensitive to it (as people can be to any number of foods—a food sensitivity is not an allergic reaction and is hard to diagnose), no definitive connection between the symptoms and MSG has ever been established by scientists, despite efforts to do so since 1968. No one can be allergic to amino acids such as MSG because they are naturally present in the human body. It's also considered generally safe.

Apropos of wondering if MSG is bad for you, consider this: Instead of sprinkling the white powder on your food, you can achieve the same effect by sprinkling food with grated Parmesan cheese, tomato sauce, or soy sauce, or by spreading Marmite yeast extract instead. All of these ingredients are loaded with the common and essential amino acid called glutamate, the salt of glutamic acid. There's a good reason spaghetti with tomato sauce and cheese is so beloved.

Glutamates are naturally present in discernable quantity in ripe tomatoes, broccoli, mushrooms (that's why they taste meaty), fish, cheese (and caseinates [p. 40]), yogurt, red wine (actually, in anything fermented), meat, and seaweed. In fact, seaweed is the original source that led to the discovery of glutamates. Kikunae Ikeda, a chemist, noted that the savory, meaty, and ultimately satisfying taste of his

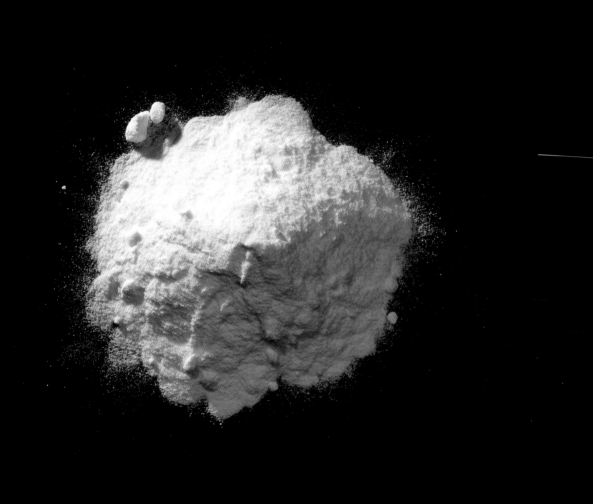

Monocalcium phosphate

$CaH_2P_2O_8$

CFR number
182.1217

E number
E341(i)

CAS number
7758-23-8

Synonyms or siblings
Calcium hydrogen phosphate, Tricalcium phosphate

Function
Process and Prep- Leavening Agent, Stabilizer
*Preservative-*Antioxidant

Description

The third basic ingredient in baking powder, after baking soda [p. 18] and sodium acid pyrophosphate [p. 152], monocalcium phosphate is just a simple blend of calcium and phosphoric acid. It's basically just dry phosphoric acid, put into baking powder (or any leavening mix) so that it will react with the baking soda to create a mad fizz when it's introduced to water or milk. It can also be used to add the tart tang to dry beverage mixes.

Eben Norton Horsford, the inventor of baking powder and a former Harvard chemistry professor, experimented for years with hundreds of calcium phosphate sources in order to create a powdered acid to mix with baking soda. The first reliable source (marketed in 1859) was animal bones from local slaughterhouses soaked in sulfuric acid. Good idea, nasty process.

Happily, a cleaner calcium source—limestone—presented itself as time moved on. Mined in the Midwest mostly for industrial use, calcium lime plays essential roles in construction, steelmaking, water treatment, and a host of other industries. Cooked at high heat into calcium oxide, or quicklime, the highly reactive stone (it burns if it gets wet) is shipped in sealed boxcars to the baking powder plant. Mixed with phosphoric acid in an intense reaction, the resulting thick liquid is dried slowly into the small, white crystals. They'll be mixed with sodium acid pyrophosphate and baking soda, and packaged into those little cans of baking powder we know and love.

Mono- and diglycerides

CFR number
184.1505

E number
E471

CAS number
67701-32-0

Synonyms or siblings
1,2,3-Propanetriol octadecanoate, Glyceryl mono- and distearate

Function
Process and Prep-Emulsifier, Stabilizer

$C_{39}H_{76}O_5$
Glycerol distearate

$C_{21}H_{42}O_4$
Glycerol monostearate

diglyceride

monoglyceride

Description

When chefs make a rich cream sauce, they may think they're using butter and cream, but in fact they're using monoglycerides and diglycerides—a special type of fat—merely in the form of butter and cream. Their close cousins, triglycerides, are found in all fats. All three have a glycerin molecule attached to one (mono), two (di), or three (tri) fatty acids. Even though they are fats themselves, the glycerides' job is to tie fat and water together. And that they do very, very well. Mono- and diglycerides are now the most widely used synthetic emulsifiers in the world.

Mono- and diglycerides are made from natural fat in two stages, and until fairly recently—they've been a popular baking ingredient since the 1920s due to their low price, ease of use, and long shelf life—they were often made from solid beef fat (tallow) or pork fat (lard). These days the starting fat is usually some kind of vegetable oil and is a result of the soapmaking process, which begins with hydrolyzing and fractionating. The fat or oil is smacked with superheated, superpressurized water ("hydrolyzed") in order to separate it ("fractionate") into fatty acids for making into soap or stearic acid, and glycerin. This pure glycerin is mixed with vegetable oil under high heat along with an alkali—usually some form of sodium—for a half hour. The result is a slick slurry, often called a milkshake, of mono- and diglycerides; its official chemical name is glycerol monostearate.

Factories dry this into flakes, powders, or beads, or ship it in the milkshake form to their food-manufacturer clients. The beads are made by spraying the milkshake slurry into a chiller vat, while flakes and powder are made by pouring the hot, oily liquid onto a cold rotating drum, forming a waxy skin that is then scraped off and broken up.

Mono- and diglycerides go to work in batter by creating small, uniform air cells and linking the water and fat. This stabilizes the batter and creates the fine grain and soft, long-lasting insides that consumers and manufacturers expect in commercial baked goods. In this way, mono- and diglycerides also postpone staling and generally reduce the need for more fat in the recipe. They are so potent that they usually account for less than half a percent by weight in a recipe.

Beyond baked goods, these emulsifying fats are used in butter substitutes, icing, peanut butter, and chocolate candy. They prevent clumping in artificial coffee creamer, where they are often paired with other emulsifiers such as lecithin [p. 168], sodium stearoyl lactylate [p. 160] and polysorbate 60 [p. 126]. They reduce the need for actual cream in ice cream by making it extra smooth. When a food company is mass-producing goodies, the choice between handling truckloads of perishable, expensive cream (no matter how yummy) and handling buckets of a stable, potent, inexpensive powder is easy.

Milk thistle extract

CAS number
84604-20-6

Synonyms or siblings
Silybum marianum, Cardus marianus, Marian thistle, Mary thistle, Holy thistle

Function
Nutrient-Fortifying Agent

$C_{25}H_{22}O_{10}$
Silybin A

Description

Few ingredients provoke such extreme statements or claims as milk thistle extract, and it is not even a regular food additive—it is a nutritional supplement. Examples of the extreme range of perspective on milk thistle from a few labels and medical institution publications might give anyone pause before using it, such as "has been used for 2,000 years as an herbal remedy for a variety of ailments." And "scientific studies show mixed results.... Problems in the design of the studies ... make it hard to draw any real conclusions." The problematic inconsistency is common for nutritional supplements because they are not required to have the extensive testing of food additives or medicines. Quite often the studies are not scholarly, well designed, or widely published.

Adding to the problem is that most gardeners think of milk thistle as a pesky weed, and at least one report notes that it has been claimed as a traditional antidote to death cap mushroom poisoning—neither characteristic likely to make it beloved among consumers today.

Regardless, milk thistle extract remains a popular herbal remedy for liver and gallbladder ailments in Europe that is catching on the United States. Milk thistle, a flowering herb related to the daisy and ragweed families native to the European side of the Mediterranean area, is now grown all over the world.

What is particularly nice about milk thistle, also called holy thistle, St. Mary's thistle, and often confused with other thistle plants, is that it attracts butterflies. When harvesting it, it's important to wear heavy gloves and protective clothing. It is a thistle, after all, and like its aggressive cousin burdock, it is difficult to handle despite its deceptively pretty red-puple flowers. The young flower heads, or leaves, are good steamed or in salads; the root can be cooked; and the seeds can be roasted for use as a relatively weak coffee substitute. It is one of the bestselling herbs in the world.

The essence of milk thistle is the silymarin extracted from its seeds. This biologically active substance is made into powders and tinctures, as well as extracts. Why it is included in energy drinks is unclear, unless it is the belief of food producers that some extra help for their consumers' livers is warranted. Silymarin, a term derived from the more accurate Silybum marianum, is actually a group of flavenoids (silibinin, silidanin, and silicritin) and is both an anti-inflammatory and an antioxidant. That is all well and good, but labs and agencies haven't said much about health benefits. There just doesn't seem to be enough information to proclaim holy thistle the holy grail of liver health.

Lycopene

$C_{40}H_{56}$

CFR number
73.5850

E number
E160d

CAS number
502-65-8

Synonyms or siblings
γ,γ-Carotene, All-trans lycopene, Tomato, Blakeslea

Function
Appeal-Color

Description

Every natural color has its most common, well-known food source, but bright red lycopene outdoes the others: it is almost exclusively linked with tomatoes, even though it is also found in paprika, pink grapefruit, guava, watermelon, asparagus (yes, asparagus), and rose hips (as well as a fungus, Blakeslea trispora). Interestingly enough, neither cherries nor strawberries have it; many red fruits are colored by anthocyanins [p. 10]. Only lycopene extracted from tomatoes is allowed to be used as a food colorant in the United States. Generally this extraction involves a large chemical plant and soaking tomato pomace left over from making juice or sauce in a volatile solvent that is boiled off (chemically synthesized lycopene is acceptable as a nutritional supplement). Lycopene is a carotene, and as such shares a molecular formula and the E-number 160 (though with a differentiating letter) with the other carotenes, especially beta-carotene [p. 20].

A phytochemical, lycopene's biological role involves helping photosynthesis along and creating beta-carotene. It joins a list of other natural pigments that are also good for you, including chlorophyll [p. 46], beta-carotene, and anthocyanins. In fact, despite the health benefits being quite unclear, many people make a special effort to eat cooked tomatoes in order to consume as much lycopene as they can.

One of the most wonderful discoveries of food science was inspired by research into the Mediterranean diet in the early 1950s. Scientists noticed that many of the residents of a southern Italian village led long and healthy lives without much need for medical care. Among the many healthful things they ate (like wine, lots of fruit and veggies, and very little meat) were tomatoes cooked in olive oil. Now we have come to understand that lycopene is only soluble, and thus made more nutritious to us, in oil (or solvents). Raw tomatoes offer much less nutrient bioavailability. Without really knowing it, the Italian villagers had chosen the best way to make tomatoes' nutrients available through something akin to natural selection. (As of the 2000s, residents in the same area are not doing as well, attributed in part to their adoption of a modern sedentary lifestyle and overprocessed-food eating habits.)

That urban legend thing is big, but silly. Starting around 1968, there were reports that someone (literally just one person) had had a bad reaction to Chinese restaurant food, with symptoms such as flushing, numbness, and a headache. The reaction to the reports ballooned, and came to be called "Chinese Restaurant Syndrome."

favorite dried kombu seaweed broth was not represented by the favorite four (sweet, sour, salty, and bitter) that were considered the only tastes. Ikeda set about to not only identify this taste chemically, which he labeled "umami," but also isolate the protein responsible for it. He achieved this goal within a year and started producing a glutamate salt in 1909. He founded a company, Ajinomoto, that currently does more than $10 billion in business around the world.

Most of Ajinomoto's amino acid products are essential items for fertilizers and feeds used in industrial farming. However, their glutamates are also processed into a white powdered flavor enhancer for cooking and tabletop use under several brand names, including the well-known Accent. It adds the umami flavor and intensifies and adds depth to existing flavors. MSG is often used in processed foods along with disodium inosinate [p. 78] because they work synergistically to great effect.

Glutamic acid and its salts, glutamates, are in yeast extract such as Marmite, so it is no stretch to note that they are also listed on ingredient labels, albeit under different names: autolyzed yeast [p. 14], hydrolyzed yeast, yeast extract, and hydrolyzed vegetable proteins. Even soy protein isolate [p. 170] has some glutamates. These are often found in salad dressing, snack foods, frozen entrées, and soups. Glutamates bump the flavor of the ingredients up a notch, and that's all they do.

Most of these glutamate products come out of Ajinomoto's Eddyville, Iowa, monosodium glutamate production facility. The process uses good old Iowa corn and a standard fermentation procedure not unlike the one used to make yogurt or beer. Bacteria is grown on sugar or molasses and starch to create amino acids that are protein building blocks, mostly used in animal feed. The Iowa processing plant was recently expanded by 50 percent, so apparently pigs, cows, and we humans don't get all that many headaches from MSG. After all, it's a natural ingredient.

Neohesperidin dihydrochalcone (NHDC)

$C_{28}H_{36}O_{15}$

E number
E959

CAS number
20702-77-6

Synonyms or siblings
Neohesperidine DC, Neohesperidin dihydrochalcone, Flavonoid dihydrochalcone, Neo-DHC

Function
Appeal-Sweetener

Description

Neohesperidin dihydrochalcone (NHDC) could be called the stealth sweetener. It functions and was discovered as part of something larger, and it disappears without a trace.

NHDC works, at absurdly low dosages (1ppm to 5ppm), as a flavor enhancer in tandem with other artificial sweeteners such as acesulfame potassium [p. 2]. It also improves the mouthfeel of the food these sweeteners are in, in part, by lingering in the mouth after ingestion. Food manufacturers like it because that synergistic talent helps lower costs. It is also stable alone on the shelf, and in foods both acidic and basic. It does all this without being absorbed by our bodies. No absorption, no calories. Food companies that use it can label their products as "no sugar added."

That mouthfeel talent is why NHDC is found in yogurt and ice cream as well as bitter medicines and toothpaste. NHDC is commonly found in chewing gum, candy, cheese, snack foods, soft and hard drinks, condiments, a wide range of desserts (especially chocolate ones), and mixed in with tabletop artificial sweeteners. Animal feed companies, particularly those with products aimed at young animals, use it to sweeten the feed.

For something so seemingly artificial, it is surprising to know that this white-yellow powder starts out as a plant extract. It's taken by solvent from the rind of bitter oranges (mostly in China) and then hydrogenated.

Despite this highly technical processing, it was actually discovered by mistake. Back in the 1960s, the US Department of Agriculture had a research program tasked with finding ways to minimize the bitter flavors of citrus juices. Coincidently, someone discovered that if one compound, neohesperidin, was hit with a strong alkali, such as potassium hydroxide, it became a super-sweetener—340 times sweeter than table sugar by weight and 1,500 to 1,800 times as sweet measured directly. It was a sweet mistake, because NHDC could be used for more than just making lemonade out of lemons.

Niacin

$C_6H_5NO_2$

CFR number
184.1530

E number
E375

CAS number
59-67-6

Synonyms or siblings
Vitamin B3, Nicotinic acid, Bionic

Function
Nutrient-Enriching Agent

Description

Niacin, also known as vitamin B3, is not only essential for growth and energy, but the other B vitamins actually cannot function properly in the human body without it. A niacin deficiency can cause pellagra, a once-common disease with horrific symptoms that include (but are not limited to) hair loss, extreme light sensitivity, skin lesions, diarrhea, dementia, and, ultimately, death. That's why niacin is always included in a white, enriched flour blend. Since adding it to the blend, pellagra has been virtually wiped out, though it should be noted that some of that success is due to more people eating balanced diets.

Niacin is one of the few vitamins that the body can make (it does this by converting the amino acid tryptophan, commonly found in fish, lean meat, whole grains, and, of course, turkey). However, we can't possibly make enough of it on our own to fight pellagra. For this we have major international chemical companies to take up the slack.

Most of the world's niacin supply is made just a bit north of the Matterhorn, not far from Zermatt, Switzerland, in a little Alpine valley town known for its old houses with flower-filled window boxes—and its ultraclean, liquified petroleum gas/naphtha cracker and petrochemical plant.

Water and air are two of the three basic ingredients of niacin. The third is petroleum, in the form of naphtha or liquid petroleum gas, which is culled from the Middle East or the North Sea and then processed by French and Italian refineries.

First, the petroleum is cracked (processed under extreme heat and pressure) into methane (which leads to acetylene), ethylene (which goes on to make the common plastic polyethylene as well as a zillion other things), and hydrogen.

In a separate process, and with an artful combination of high and low pressure and heat exchangers, air is liquefied and separated into nitrogen and oxygen to make ammonia, which is eventually mixed with a little hydrogen from natural gas to make nitric acid. This nitric acid is the source of the nitrogen in niacin, which is significant because the discovery of the nitrogen part of the niacin molecule—the "amine"—led to the discovery of all vitamins. In fact, the word "vitamin" is a corruption of the chemical term, "vital amine," the term used first by its discoverers.

Next, the ethylene and acetylene are mixed under pressure with some water and a rare platinum catalyst to make acetaldehyde, a flammable liquid, which is then mixed with the ammonia that was made earlier in the process. This ammonia/acetaldehyde blend is then mixed with the previously produced nitric acid to create a white solid block: pure niacin. The block is milled into the familiar flourlike powder that we use in our food and pills. It is a long, complex, and wholly unnatural route from petrochemicals to an absolutely essential health ingredient.

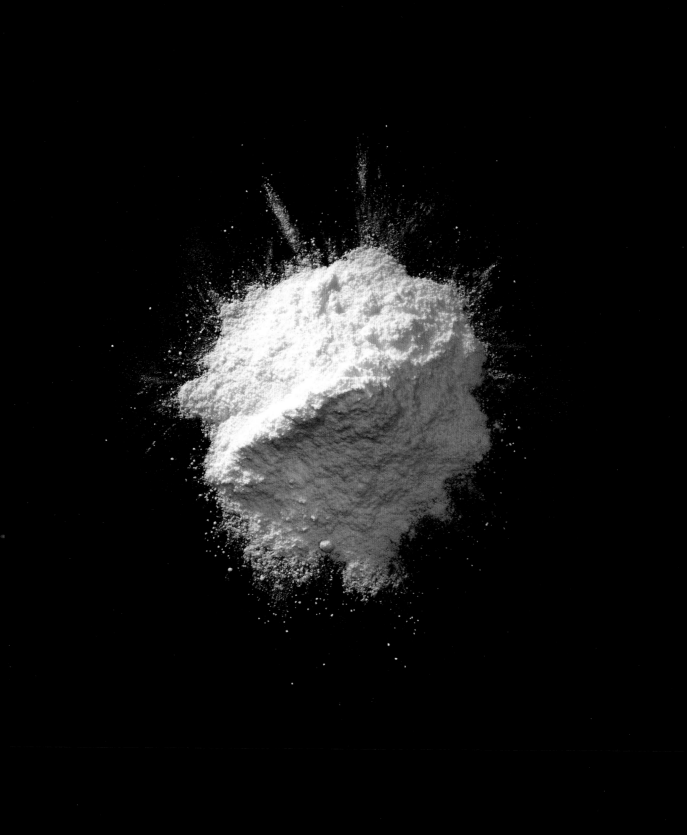

Phosphoric acid

H_3PO_4

CFR number
182.1073

E number
E338

CAS number
7664-38-2

Synonyms or siblings
Orthophosphoric acid, Phosphoric(V) acid, Trihydroxylphosphine oxide

Function
Preservative-Antimicrobial, Antioxidant

Description

Phosphorus is one of the seven elements necessary for life, the atomic bonder of the amino acid rungs in your DNA. It is also so flammable that it bursts into flame when it comes in contact with air. When turned into phosphoric acid, it becomes one of the most popular food additives in the world. It puts the tang into the taste of Coca-Cola, emulsifies processed cheese, and makes baking powder fizz in any baked food product (see the suffix "phosphate" in the list of ingredients). It sets jelly and chocolate pudding, gels processed seafood (surimi, or sea legs), preserves meat in sausage, refines sugar, and much more, almost always as less than 1 percent of the food product.

It also is part of fertilizers, fire retardants, and, in the form of Naval Jelly, a great rust dissolver. The reason both Coke and Naval Jelly are good at removing rust is that they both have phosphoric acid in them, though in vastly different concentrations.

Phosphorus and phosphoric acid have been around for a while. An amateur German alchemist became the first known discoverer of any element when he isolated phosphorus in the mid-1600s by distilling urine he collected from nuns (he figured theirs would be the purest around). These days phosphorus is recovered from less exalted sources, such as various phosphate ores around the world.

Most industrial phosphoric acid is the result of the "wet process" technique of reacting phosphate rock with sulfuric acid to instantly create phosphoric acid along with a by-product of gypsum. The reaction is so vigorous that it takes place in what's called the "attack tank." Phosphoric acid is the seventh most common industrial chemical produced in the United States, used in soaps, detergents, water treatment, and a host of reactions for various chemicals.

Elemental phosphorus is produced in Soda Springs, Idaho, where the soft, dirtlike ore is roasted with silica (sand) and coke (pure carbon from coal) in an electric arc furnace at 11,000 degrees Fahrenheit, as hot as the sun's surface, producing a plasma that cools into a liquid and then solidifies into a soft wax. That phosphorus is sent elsewhere to be burned into a gas. When sprayed with water, the reaction creates pure, liquid phosphoric acid—most of which is used to make Monsanto's Roundup herbicide, and due to its purity, food products such as baking powder and cheese. Pure phosphoric acid is thus used to both kill things and give life, making it versatile indeed.

Polydimethylsiloxane (PDMS)

$(C_2H_6OSi)n$

CFR number
173.340

E number
E900

CAS number
63148-62-9

Synonyms or siblings
Dimethylpolysiloxane, PDMS, Dimethicone

Function
Process and Prep- Emulsifier, Powder Flow Agent, Stabilizer

Description

Wonderful things happen when you mix borax with silicone oil, as some researchers learned while searching for an alternative to rubber during the depths of World War II. Their discovery turned out to be a great lubricant and food additive, among myriad other things. They also noticed that it bounced and stretched and was fun to play with. Yes, these scientists actually discovered Silly Putty! And formulating it with just a few different ingredients in a giant chemical plant makes it into a great antifoaming agent: polydimethylsiloxane (PDMS).

PDMS is a minor and peripheral additive, but it makes its way onto food ingredient lists for a very important reason; only a tiny bit—10ppm—keeps hot deep-fryer oil from splattering when frozen or when moist food gets dropped into it. This important safety and hygiene feature also prolongs the life of frying fat, something well-appreciated by the major fast food and casual food chains.

Mixed into the oil by manufacturers, PDMS barely registers in food itself; it is so nontoxic, nondigestible, tasteless, odorless, and nonreactive that it is also used to clean syringes and make contact lenses. Speaking of nontoxic, Silly Putty, being a toy, must be nontoxic. Certainly, if it had been proven toxic, it would pose a dramatic problem for kids, what with more than 300 million pieces sold since 1950 (currently they sell 9 million "eggs" each year). In fact, it is actually used as a skin protectant in concentrations up to 15 percent.

It is also used as an antifoaming agent in soft drinks, instant coffee, chewing gum, vinegar, sports and energy drinks, sweet snacks, syrups, and chocolates. It plays a role in making skim milk and fermenting wine. A sibling product, simethicone, is popular with fermenters and detergent manufacturers for their issues with foam control, as both products make lots of bubbles. More commonly it is used in familiar over-the-counter products that relieve stomach gas pain, such as Maalox, Mylanta, and Gas-X. The antifoaming mechanism weakens the gas bubbles so that they combine in ways that make it easier to get them out of your system. It makes you feel a lot better and, because it is not digestible, it passes right out of your body.

That long name only describes the molecule, which has repeating ("poly") groups of two ("di") methyl molecules made of carbon and hydrogen, along with silicon and oxygen ("silox") and certain carbon-hydrogen combinations ("ane"). It can also be referred to as dimethylpolysiloxane or dimethicone, though the latter name is more often associated with a version made for cosmetics. It is the stuff that keeps your hair slippery. Some vendors just refer to it as silicone oil, not even bothering with the actual facts of manufacturing. Keeps you on your toes.

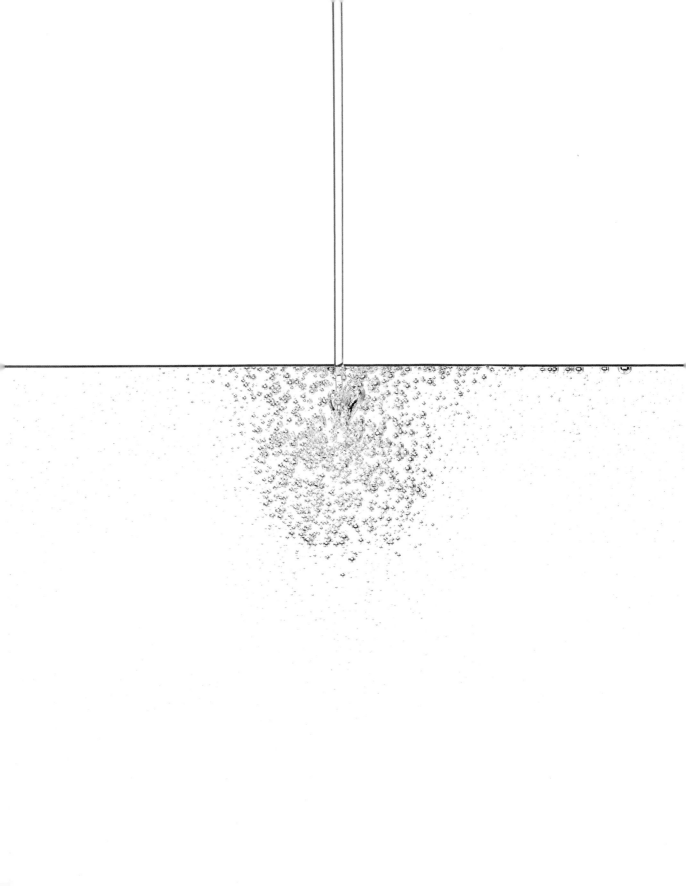

Polyglycerol polyricinoleate (PGPR)

$(C_3H_5O_2)n(C_{18}H_{32}O_2)m$

CAS number
29894-35-7

Synonyms or siblings
PGPR, Polyglycerol polyricinoleic acid

Function
Process and Prep-
Emulsifier, Stabilizer

Description

Leave it to modern food science to turn something so reviled as castor oil—once known as a crude laxative—into something so beloved as chocolate. This miracle of science is done by using a process that relies on basic soapmaking techniques.

Superheated glycerol—the fatty acids separated out of vegetable oil in the refining and high-heat fractionating process that supplies glycerin to the soap industry—turns into polyglycerol, which is mixed with a polymer of castor oil fatty acids that has also been refined and fractionated. The resultant light yellow syrup has such a long name, polyglycerol polyricinoleate (based on its two subingredients), that it is usually referred to only by its acronym, PGPR.

Castor oil and its derivatives are used in a number of interesting ways. They can be found in antifungal products, skin creams that treat acne and eczema, analgesic wound dressings, eyedrops, and arthritis treatments. Castor oil is popular with some hair repair products and has industrial uses, too: it's featured in the production of plastic, rubber, adhesives, paints, and lubricants.

Glycerin is a well-known ingredient in medicines as well as foods, though more often as a subingredient (as it is here) in things like mono- and diglycerides [p.104]. Cooking the heck out of castor oil and glycerin and combining them creates a superb emulsifier that is particularly helpful to chocolate makers. They rely on PGPR through thick and thin, so to speak. Thick, as in a way to reduce the use of the more expensive cocoa butter in chocolate bars, and thin, as in controlling the viscosity in thin chocolate coatings on candy. In both cases it almost always must work alongside the old standby, lecithin [p.168]. It also adds some stability to a recipe involving a natural product, chocolate, that might vary in some characteristics from batch to batch. Chocolate manufacturers depend on it.

Cooking the heck out of castor oil and glycerin and combining them creates a superb emulsifier that is particularly helpful to chocolate makers. They rely on PGPR through thick and thin, so to speak.

Chocolate products must have an alluring mouthfeel. The gritty chocolate must feel smooth, and the fats and oils must not separate. In high-speed manufacturing lines, the molten chocolate has to flow just right through the machinery without breaking down. Emulsifiers are needed to tie fat and moisture together. Invented in the 1950s, PGPR's toxicological testing and government approval took decades. It was approved in 1995 in Europe and about ten years later in the United States. Since about 2006, some of the best-known, mass-market brands of chocolate have used just a touch of PGPR in their recipes—only between 0.1 percent and 0.3 percent.

In addition to facilitating the magic of chocolate making, PGPR works as a great emulsifier in spreads, salad dressings, and margarines. Not a bad repetoire for something made from such unlikely subingredients.

Probiotics

CAS number
308084-36-8

Synonyms or siblings
Lactobacillus, Acidophilus, Acidophilus bifidus, Acidophilus lactobacillus

Description

Probiotics are the opposite of antibiotics, and you need them in your body—they are present in your intestinal system. They are called your gut flora, and your belly is full of them. They may be made from stuff in your body, but you can easily add more to your system by eating familiar foods.

Probiotics are live microorganisms such as bacteria or yeast that, in adequate amounts, we think benefit our health. More than 500 kinds of "good" bacteria live in our digestive systems, working hard to keep us healthy. They perform basic functions such as digesting food and helping maintain our immune systems.

In the early 2000s, food product marketers put probiotics on a pedestal with organic food and other highly salable food and nutrition trends and the market for them exploded. Though there are relatively few studies, some work done in the 1980s and 1990s indicated that probiotic supplements could help fight digestive disorders such as rotaviral diarrhea and lactose intolerance. Friendly organisms, the thinking goes, can improve intestinal function that may have been disrupted by a bacterial imbalance, intestinal condition, vaginal infection, a treatment of antibiotics, or surgical (physical) damage. (Check with your doctor before taking any probiotics if you are pregnant, nursing, have intestinal damage, immune problems, or any other worrisome condition. They can cause problems or may be mislabeled.)

Despite these initial reports, medical science in general is not all that excited about probiotics for a number of reasons, most of which stem from a lack of sufficient and thorough studies. Don't hold your breath for more scientific research. Because probiotics are a supplement, they aren't regulated or approved by the FDA and thus manufacturers are not likely to pay for costly and time-consuming studies. Still, doctors can and do prescribe specific strains of probiotics for specific health issues. Probiotics do appear to be part of a healthy diet when they are ingested in their natural form.

And what is that elusive-sounding natural form? Yogurt—as long as it is labeled to contain live and active cultures (lactobacillus acidophilus is the most common; these are found in your mouth, intestine, and vagina)—kefir, and soft fermented cheese. You can also find them in most fermented

In order to grow probiotics, scientists can harvest as little as one cell of the purified bacteria. They must take exquisite care to properly feed the cells and control their environment and fermentation process.

food: kimchi, pickles made without vinegar, unpasteurized sauerkraut, buttermilk, sourdough bread, miso, and tempeh. The consumer searching for a particular bacteria and dose has to rely on a precisely labeled supplement or food product, such as "acidophilus milk."

If you want an even bigger boost for your bacteria buck, eat foods known as prebiotic—they feed the probiotic bacteria. Good examples of prebiotics are asparagus, bananas, oatmeal, honey, many legumes, and, happily, red wine. Some foods, called symbiotics, naturally combine both pre- and probiotics. Yogurt and kefir are the best examples of symbiotics.

Manufacturing probiotics is simple but costly, which is why there are only a few major manufacturers in the world. The three biggest companies got their start in the dairy business and also make yeast. This expensive process begins in a laboratory, where desired microorganisms are harvested from living sources (animal or plant), identified, and isolated. The good news is that the microorganisms are naturally self-replicating—all you need to do is feed them.

In order to grow probiotics, scientists can harvest as little as one cell of the purified bacteria. They must take exquisite care to properly feed the cells and control their environment and fermentation process. The lab starts with 1 milliliter of "seed" bacteria in the bottom of a test tube frozen in liquid nitrogen at -112 degrees Fahrenheit. Days later, after a series of controlled overnight fermentations in a perfectly clean and highly computerized plant, that seed's offspring will fill a 30,000-liter, two-story-high stainless steel tank.

The final liquid result can be freeze-dried, powdered, and sent to clients in the animal nutrition, winemaking, baked good, and, of course, nutritional supplement industries. Supplements are blended with neutral ingredients—rice maltodextrin is popular—for packaging as capsules at the desired strength, and the natural stuff becomes a processed item.

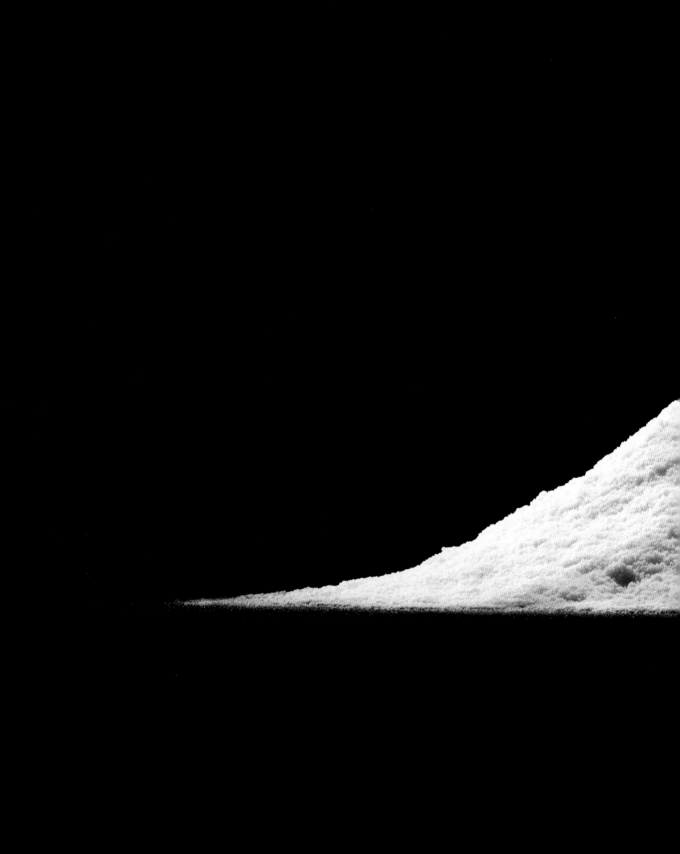

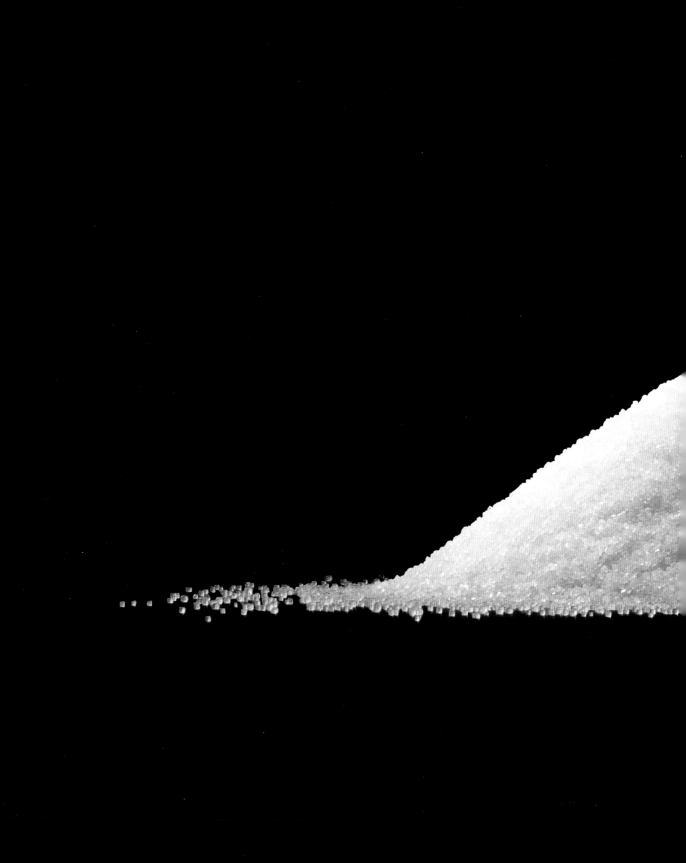

Salt played a major role in world trade among the ancient Greeks, Romans, and Chinese. Wars were fought over it, and it was once as valuable as gold (early Chinese coins are said to have been made of it, and Roman soldiers may have been paid in it, giving rise to the word "salary").

While salt is also one of the basic tastes (salt, sweet, bitter, sour, and umami) and necessary in our diet—our cells need it in order to function—it is used extensively simply to enhance flavors, even sweet ones like chocolate. That's why your chocolate fudge and cake recipes almost always call for a pinch of salt. Salt is a functional ingredient, classified as a "processing aid," which means it enhances not only the flavor but also the texture. Salt works as a chemical buffer to help "bind" the dough, controlling the yeast/sugar fermentation action and allowing the batter to hold more water and carbon dioxide so that it can expand more easily. This cuts down on the big "voids," or bubbles, to create a uniform grain, texture, and strength.

Salt helps in as many as 16 other essential food functions (one source at Morton says 21): it fights bacteria and mold growth (a good bacteriostat, salt water is a traditional wound cleanser), activates and "sets" food coloring (think red hot dogs and hams; it is used extensively in textile dyeing, too), creates texture and rinds for cheese, acts as a binder for sausages, and, of course, preserves everything from pickles to fish. Salt accomplishes this last task by absorbing moisture from bacteria and mold through osmosis, killing the cells or at least preventing them from reproducing. Salt can't be considered just a food product, though. Because it is a basic chemical and a source of sodium or chlorine for making other chemicals, it has an estimated 14,000 industrial uses.

All of this is such a far cry from mere taste that it seems like salt doesn't get enough credit. Remember that next time someone asks you to pass the salt.

Salt

NaCl

CFR number
100.155

CAS number
7647-14-5

Synonyms or siblings
Sodium, Sodium chloride

Function
Appeal-Flavor Enhancer

Na^+ Cl^-

Description

We've been preserving food with salt, the most common and oldest-known food additive, for millennia. Early human settlements were often made near salt outcroppings, which attracted animals and, therefore, our ancestors, who were interested in hunting and eating those animals. Our earliest ancestors salted meat, fish, and vegetables—including olives—to preserve them and to satisfy their healthy craving for salt. Salt played a major role in world trade among the ancient Greeks, Romans, and Chinese. Wars were fought over it, and it was once as valuable as gold (early Chinese coins are said to have been made of it, and Roman soldiers may have been paid in it, giving rise to the word "salary"). Genoa and Venice rose to prominence as centers of the salt trade. Through the centuries, the message has been clear: we need salt. More than 200 million tons of it are mined each year, worldwide.

There are three main ways to harvest salt, yet another edible rock. You can evaporate salt water to get crystals naturally—the oldest method—but that takes consistently sunny, windy weather found only in limited (but beautiful) places. You can blast it out of a mine with dynamite and front-end loaders and then crush it, but that takes big mining operations and leaves a fairly impure rock salt suitable for water softening, ice control, and chemical processing, but not for food. Or you can flush it out of the ground with water and boil it down in an evaporation plant.

This last method, called solution mining, produces almost all the fine, pure crystals of salt used both at home and in processed foods. Table salt is actually pumped out of the ground as brine from U-shaped wells (water is pumped way down into deep salt deposits and the resultant brine flows up) and piped over to a processing facility, where the brine is concentrated and dried. It is also ground into various smaller crystal sizes for different uses, and it is often mixed with an additive such as iodine for health reasons, or calcium silicate to help it flow. That last reason is the genesis of the Morton Salt motto, "When It Rains, It Pours," in use since 1914.

Riboflavin

$C_{17}H_{20}N_4O_6$

CFR number
184.1695

E number
E101

CAS number
83-88-5

Synonyms or siblings
Vitamin B2, Lactoflavin, 5'-Phosphate sodium

Function
Appeal-Color
Nutrient-Enriching Agent

Description

Riboflavin, or vitamin B2, is another essential. Without it, we'd have trouble growing. We'd suffer cracks around the mouth, sores around the nose and ears, a sore tongue, and light-sensitive eyes. Most importantly, we would fail to convert food into energy—more than enough reason to include it in the white-flour enrichment mix. It's deep orange-yellow, a color so intense that it is also used as a food coloring, especially for Easter eggs.

Of course, the coloring can be a problem. It's extremely hard to get out of your clothes if you happen to brush some on you, and if you take extra vitamin B2 supplements your urine turns an especially bright yellow. One manufacturer had to build a treatment plant just to get rid of the orange in its wastewater. That's intense.

Speaking of manufacturers, most are in China. While some use chemical synthesis to make B2, most fermented it from microorganisms found in yeast, fungus, or bacteria. Candida yeasts are common; Ashbya gossypii fungus is used to make about 30 percent of the world's supply of B2, while other large producers favor a bacteria called Bacillus subtilis. Some make it from spent beer grain, recycled by the beer companies. In nature, B2 comes from leafy green veggies, liver, fish, milk, and poultry.

Generally, Chinese vitamin companies ferment riboflavin by incorporating their yeast, fungus, or bacteria (what they call the "master organism") in a stew of various fats or carbohydrates. Temperature and nutrients are combined in ways that vary from place to place as much as recipes vary from chef to chef. It might be a stinky mix of nutrient-rich waste fats, or cod liver oil or canola or soybean oil. Some companies use a carbohydrate mash made of sugar from beets, cane molasses, or liquid rice; glucose from corn is another popular choice. Other manufacturers use specially treated millet seeds, kept for a week at the optimum breeding temperature of 90 degrees Fahrenheit. The enzymes that live secrete what becomes riboflavin.

The vitamin is extracted from the fermentation broth through a complex process that involves multiple steps (concentration, purification, crystallization, drying, and milling) in order to obtain a deep-orange flourlike powder that smells faintly like rotten wood, an inauspicious beginning for an essential nutrient.

Red No. 40 and Yellow No. 5

$C_{18}H_{14}N_2Na_2O_8S_2$ and $C_{16}H_9N_4O_9S_2Na$

CFR number
74.340 (Red No. 40)
74.705 (Yellow No. 5)

CAS number
25956-17-6 (Red No. 40)
1934-21-0 (Yellow No. 5)

Function
Appeal-Color

E number
E129 (Red No. 40)
E102 (Yellow No. 5)

Synonyms or siblings
Allura red AC, Allurar red, FD&C Red No. 40 (Red No. 40)
Tartrazine, Lemon yellow, FD&C Yellow No. 5 (Yellow No. 5)

Red No. 40

Yellow No. 5

Description

One would think that a red or yellow dye would be made from red or yellow ingredients, but then one would be wrong. Red No. 40, also known as allura red, is among the most common colorants around. It's made from a few gray powders, one of which is nitric acid. Yellow No. 5, also known as tartrazine, is made similarly but with tartaric acid. Combined to create neutral salts akin to Epsom salts, the subingredients for most artificial colors originate as petroleum products such as naphthalene and benzene in China. Liquid acidic mixtures are neutralized with a dose of sodium hydroxide (lye) in giant stainless steel vessels called coupling tanks.

Once thoroughly reacted, mixed, filtered, and purified, colors are atomized in a spray drier with walls so hot that when the mist hits them, it dries immediately into a perfectly colored powder. The end result for Red No. 40 gets a long chemical name: 6-hydroxy-5-(2-methoxy-5-methyl-4-sulfophenylazo)-2-naphthalenesulfonic acid sodium salt. Yellow No. 5 has a molecule that looks only a bit different, 5-hydroxy-1-(p-sulfophenyl)-4-(psulfophenyl)azopyrazole-3-carboxylic acid trisodium salt. You can call them by their FDA numbers, though.

We've been coloring our food since 5000 BCE. Since at least the 1300s we've been making butter more yellow either with saffron or marigolds. These days, food coloring is added for psychological reasons, whether aesthetic or commercial.

Modern artificial colors date back to the first synthetic dye, mauve—the color discovered in 1856 by British chemist Sir William Henry Perkin. The story goes that 15-year-old Perkin accidentally discovered aniline while experimenting with derivatives of coal tar left over from burning and processing coal into coke for iron-making. He had been trying to make quinine and ended up with a dark goo that, when he attempted to clean it with alcohol, dissolved into a dark purple dye. Eventually, the textile industry snapped it up, and Perkin went on to synthesize violet and green as well as natural fragrances from aniline, giving rise to some of the most common artificial ingredients known today. Most of the early colorants were toxic; only with the increased sophistication of both chemistry and government in the 1900s were edible versions created and labeled. Constant testing has eliminated some more in recent times.

Colorants are used sparingly. A figure of 50ppm to 100 ppm is typical for cakes, for example, or only 5ppm to 10ppm in drinks such as pink lemonade. That's not a whole lot, which is why colors are usually the last items listed on an ingredient list. Despite their strength (they are used in microdoses), more than 17 million pounds of artificial food coloring are made every year in the United States. With that rate, it is clear that we really, really love our food colored.

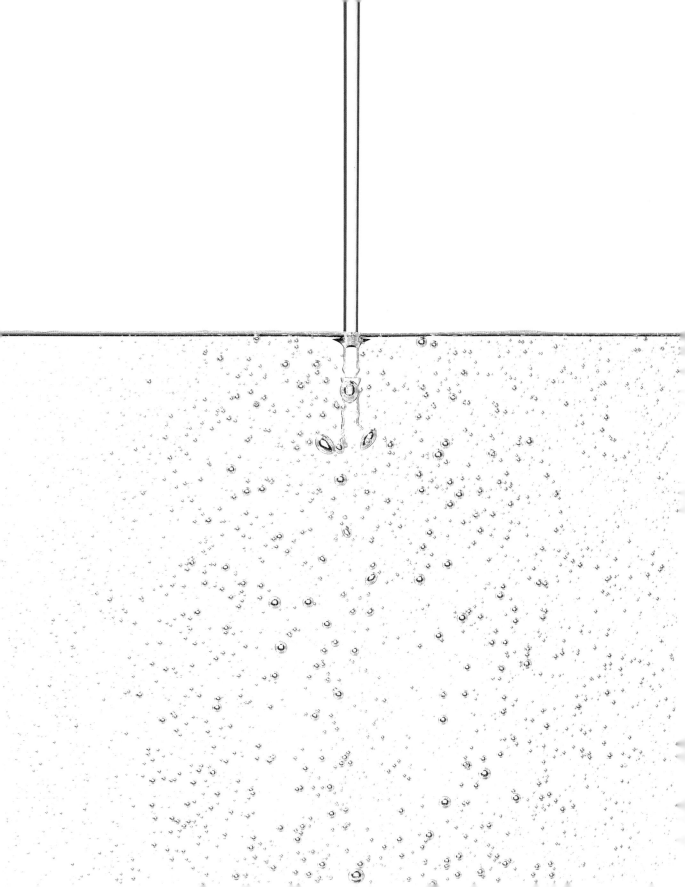

Starting from crude oil or natural gas that is cracked (blasted with high heat), propylene gas is created and then reacted with chlorine or lye to make propylene oxide. All that is needed to turn it into alcohol is to react it with that old standby, water. Despite its crude origins, PG has been used in food, animal feed, cosmetics, and pharmaceuticals for more than fifty years.

cracked (blasted with high heat), propylene gas is created and then reacted with chlorine or lye to make propylene oxide. All that is needed to turn it into alcohol is to react it with that old standby, water. Despite its crude origins, PG has been used in food, animal feed, cosmetics, and pharmaceuticals for more than fifty years.

One reason for its widespread acceptance is that it is less toxic than alcohol and limited to 0.1 percent to 0.3 percent by weight of any recipe. Once digested, it is quickly metabolized much as sugar is: converted into lactic acid and excreted in urine. Happily for a petroleum product, it is entirely biodegradable.

Propylene glycol

$C_3H_8O_2$

CFR number
184.1666

E number
E1520

CAS number
57-55-6

Synonyms or siblings
Propylene glycol,
α-Propylene glycol,
1,2-Propanediol,
1,2-Dihydroxypropane

Function
Preservative-Antioxidant
Process and Prep-
Moisture Control Agent,
Powder Flow Agent,
Stabilizer, Thickening
Agent

Description

For something that is colorless and flavorless, propylene glycol sure plays an important role in artificial colors and flavors. It is in almost all of them. Sure, it helps thicken, clarify, and stabilize beer, salad dressing, and baking mixtures, but its role in food is mostly as a solvent and carrier of artificial colors and flavors. It makes the color pigments disperse evenly in any mixture and keeps the flavor smooth and thick.

Consumers have a lot more contact with PG, as it is called. Beyond artificially coloring and flavoring food, it plays a role as a humectant—or wetting agent—and emulsifier in sunscreens, shaving creams, shampoos, personal lubricants, cosmetics, and medicines, including vaccines. But that accounts for only about a fifth of the PG produced. The rest, for a total of an astounding worldwide consumption of well over 2 billion pounds each year, goes into industrial uses as part of airplane deicers, printing ink solvent, laundry detergent, and plastics.

Being hygroscopic, it is sold alongside cigar cutters and lighters to cigar lovers as a necessary ingredient in their humidors. It's also the fuel for theatrical smoke machines. Speaking of consumer contact, it is the primary ingredient in paint balls.

As its name suggests, propylene glycol is based on petrochemicals. Starting from crude oil or natural gas that is

Propionate (calcium and sodium)

$C_6H_{10}CaO_4$ and $C_3H_5NaO_2$

CFR number
184.1221
(Calcium Propionate)
184.1784
(Sodium Propionate)

E number
E282
(Calcium Propionate)
E281
(Sodium Propionate)

CAS number
4075-81-4
(Calcium Propionate)
137-40-6
(Sodium Propionate)

Synonyms or siblings
Calcium propanoate,
Calcium dipropionate,
Mycoban
(Calcium Propionate)
Sodium propanoate,
Napropion
(Sodium Propionate)

Function
Preservative-Antimicrobial,
Shelf Life Extender

Calcium propionate

Sodium propionate

Description

Calcium propionate does two things: it's a very common mold inhibitor in bread, and it contributes a few calcium ions to help a major enzyme in bread dough called α-amylase work on the bread starches so that they can more readily react with the yeast. All that is to say that calcium propionate helps improve bread's texture while fighting mold, which is why it remains one of the most popular antifungals in all kinds of baked goods. Its sibling sodium propionate is used a bit less because it not only lacks that helpful calcium but also adds some sodium to the food—and who needs that?

These propionates are among the few additives that are already present in many foods, especially butter and Swiss cheese. The smell of its precursor chemical, propionic acid (calcium and sodium propionate are the salts of the acid), is described as acrid and offensive, which makes sense considering that it is a component of the smell of human sweat.

Propionates fight mold much in the same way sodium benzoate [p. 154] does, by disrupting the microbes' energy production, but they work in a wider variety of environments (benzoates need an acidic environment). Outside the bread aisle, calcium propionate is found in processed meat and a variety of dairy products. It can also fight fungal infections on your skin (especially the acne bacteria Propionibacterium); propionates are commonly used in cosmetics and medicines as preservatives, bactericides, and fungicides.

Though propionic (also called propanoic) acid was discovered in sugar in 1844, it was not until 1941 that the acid was first synthesized. The honor goes to a German chemist working for the major chemical firm BASF, who made it from ethyl alcohol and carbon monoxide. BASF was not able to build on its discovery right then, during World War II, but today the modern chemical plant there, in Ludwigshafen, Germany, remains one of the biggest in the world. However, even BASF makes propanoic acid in China now, as do most of its competitors. They still use the original method, though slightly modified, and make close to 170,000 tons of it each year.

When propionic acid was first introduced to industry, it was as a fungicide and bactericide to fight mold in animal feed, which is still its most common use (it is also used as a food additive by itself). Farmers spray it on their corn and grain stored in their silos. A liquid fatty acid, it breaks down into CO_2 and water; in your body, it is metabolized and eventually excreted as CO_2 via the Krebs cycle. Beyond food, propionic acid is used as a chemical building block for many pharmaceuticals, herbicides, dyes, plastics, paint, vitamin E, and (amazingly for something that smells like sweat) perfumes and flavoring agents.

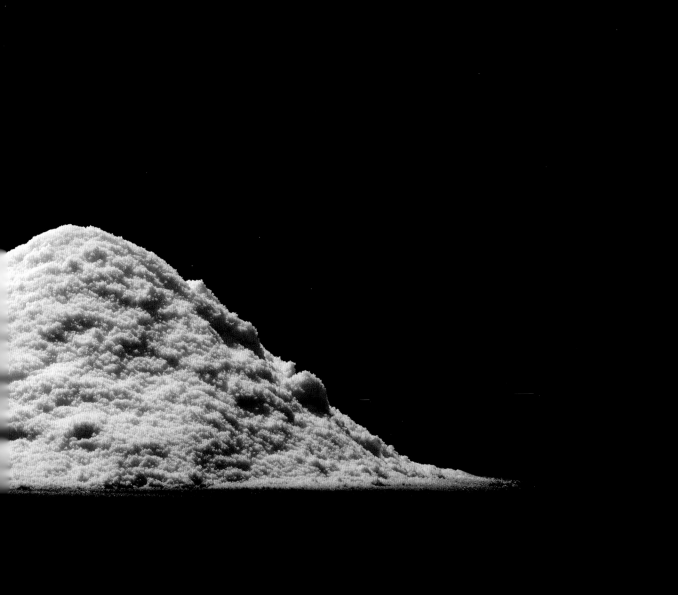

Shellac

CFR number
175.300, 73.1(b)

E number
E904

CAS number
9000-59-3

Synonyms or siblings
Gum lac, Orange shellac, Confectioner's glaze, Resinous glaze, Confectioner's resin, Lac wax, Candy glaze, Pure food glaze, Natural glaze, Lac resin

Function
*Appeal-*Color
*Preservative-*Shelf Life Extender

$(C_{31}H_{48}O_7)n$
Jalaric acid

Description

It's hard to wrap your head around the idea that shellac is what it is, mainly because its nonfood uses and its source are so improbable. For years, before vinyl was introduced in the 1940s, records were made of it, and it is best known (when mixed with denatured alcohol or sold as dry flakes) as a varnish for wood furniture, floors, and fine stringed instruments. Made from insect excretions, shellac is full of surprises.

It is a great sealant that keeps foods fresh; it keeps moisture in or out, often replacing the natural wax that is found on apples that may have been removed by washing. Shellac is also used on vegetables, chocolates, baked goods, and even coffee beans and chewing gum. It is edible, but it is used in very thin films so there is no taste or waste. Shellac coating keeps some pills and tablets, especially common aspirin, from being digested too quickly or from absorbing moisture while on the shelf. It's usually listed under the generality "enteric coated" or "pharmaceutical glaze." On food, it might be called the ambiguous "resinous glaze," "food glaze," or "natural glaze," but also the more suggestive "beetle juice," "lac resin," and even plain old "shellac."

The hardy source of this useful product is a small-scale insect, Kerria lacca (Laccifer lacca), or "lac beetle" (despite the fact that it is not a beetle). It swarms on certain kinds of trees, mostly in India, but Kerria lacca is also found in China and Thailand. Now cultivated, lac has been harvested as a varnish since at least the late 1500s. In fact, its use as a dye possibly goes back to antiquity.

In their larval stage, the lac insects secrete a hard, waterproof tube of resin as they move around and feed on the tree sap (in a way, they are merely processing the sap). After the insects exit, the tubes are scraped into canvas bags, heated, strained through the canvas, and dried into flakes or pucks for further processing with alcohol, alkalis, or other solvents. While simple, the process is hardly efficient: it takes more than 100,000 lac bugs to make just over a pound of shellac flakes. Remarkably versatile, shellac is actually a natural plastic (hence the use in making records); specifically, it is a natural bioadhesive polymer that is chemically similar to synthetic polymers.

"Gum lac" may be great on food, candy, and pills, but its varnishing abilities also make it a strong candidate for sealing wax, hair spray, mascara, nail polish, perfume, and lipstick. Industrial nonvarnish uses may include use in printing ink and slow-release fertilizer. Shellac is a skin irritant for some, but most vexing, perhaps, is that because it is made from an insect, it is not a vegetarian item. In other words, a shiny apple in a supermarket may not be suitable for vegetarians.

Silicon dioxide

SiO_2

CFR number
172.480

E number
E551

CAS number
7631-86-9

Synonyms or siblings
Quartz, Silica, Silicic oxide, Silicon(IV) oxide, Crystalline silica

Function
Process and Prep-Powder Flow Agent

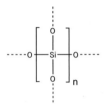

Description

For starters, this unlikely food ingredient is part of most rocks. It is made of the two most common elements in the world, literally: oxygen and silicon. You know it as sand or quartz. And while it would not be wrong to associate sand and rocks with dryness, moisture is what it is all about. Powdered silicon dioxide—it does not exist as pure silicon in nature—soaks up moisture so well that manufacturers like to add it as an anticaking agent to salt, flour, soup mixes, chips, sugar, and any other powdered food or food product that can't handle moisture well. Even when ground into a fine powder, it retains its porous structure and soaks up moisture like a sponge.

Being that it is not a food, per se, silicon dioxide is generally limited to less than 2 percent of a given food. While most of us associate silicon with the chips in our smartphones, it's common—in the form of silicon dioxide—in essential fruits and vegetables, including oranges, apples, grapes, beans, peas, cucumbers, celery, and nuts (especially peanuts and almonds). Whole grains and rice are other excellent sources of silicon. We actually need it, along with other minerals, for healthy bone development. Brewers use silicon dioxide as a beer defoamer.

We use silicon dioxide to make glass and computers. In its pure, natural powder form (as in ground-up, fossilized, hard-shelled algae-like organisms known as diatomaceous earth), we use it to control garden pests and bedbugs. In this last case, it works differently from its food and industrial uses—physically, not chemically. The sharp edges of the grains cut into hard-shelled insects' exoskeletons and cause them to dry out.

Most silicon dioxide and related compounds are used in the concrete and glass industries, or as part of the computer chip industry. Manufacturing a pure product from sand requires multiple steps in giant industrial plants. Generally this starts with heating sand and coke (cooked coal) at extreme temperatures to create gas that is then broken down into the desired components. Gases are then cooled to form liquids and solids that eventually yield pure silicon dioxide. A complex process for a simple thing: sand to sand.

Sodium acid pyrophosphate (SAPP)

$Na_2H_2P_2O_7$

CFR number
182.1087

E number
E450(i)

CAS number
7758-16-9

Synonyms or siblings
Disodium pyrophosphate, Disodium dihydrogen pyrophosphate, Disodium diphosphate

Function
Preservative-Antioxidant
Process and Prep-Emulsifier, Leavening Agent, pH Control, Powder Flow Agent

Description

Sodium acid pyrophosphate is just a simple blend of sodium and phosphoric acid and is a major component of commercial baking powder. It fizzes in reaction to oven heat and makes bubbles in batter. In other words, it provides the second action in "double-acting" baking powder. Without SAPP, there would be only one round of fizzing—when water or milk mix with the dry ingredients. Things would flatten out by the time the batter actually got hot. It's a control issue.

The inventor of baking powder, Eben Norton Horsford, a former chemistry professor at Harvard, recognized this problem. After struggling to find a solution for years, in 1885 he discovered that sodium acid phosphate gave off gas in response to heat, not water. Horsford mixed it with sodium bicarbonate to make the first phosphate-based, stable, reliable, affordable baking powder. He named his new product in honor of the great Count Rumford, whose beribboned, ponytailed cameo still graces the label on Rumford Baking Powder cans today.

SAPP manufacturing starts in the ground, with the mining of its two mineral subingredients. Phosphate ore is processed into either pure elemental phosphorus or mixed with sulfuric acid and then made into phosphoric acid [p. 116]. Truckloads of this dangerous, oily liquid are delivered to a house-size spherical tank outside the SAPP factory, known as a "salts" plant. There it sits, ready to attack.

The second mineral, sodium carbonate, is mined in Wyoming. Railroad hopper cars drop their loads onto a conveyor that takes the mineral to a 5,000-gallon reacting kettle, where it is deposited along with a good dose of the already delivered phosphoric acid. The resulting thick liquid, after passing through two different mixers, is cooked for an hour by spiraling through a 30-foot-long, 10-foot-diameter, rotating oven. The 450-degree Fahrenheit heat purifies the liquid, slowly transforming it into crystals. A ceiling-high, funnel-shaped machine called a cyclone completes the drying and crystal-sorting process with a vigorous spin cycle.

The final product is fetchingly nicknamed "pyro," which is a chemical term that means "chain of two," as in a single molecular chain formed by heat from two separate phosphate molecules (phosphorus and oxygen).

Though it is mostly used in food as part of baking powder, sodium acid pyrophosphate has other uses. It maintains color in canned seafood, stabilizes cured meats in a reaction with sodium nitrite, controls fermentation in roasted meats, and keeps frozen potato products from turning brown. It also keeps instant noodles from getting sticky and mushy. On the edges of the food industry, SAPP is useful in hog and chicken butchering, specifically as part of the soup that removes hair and feathers. Outside (way outside) of the food industry, SAPP can be found removing iron stains from leather hides and loosening up drilling muds in oil rigs, very uncakelike activities, indeed.

Sodium benzoate

$C_7H_5NaO_2$

CFR number
184.1733

E number
E211

CAS number
532-32-1

Synonyms or siblings
Benzoate of soda, Dracylic, Benzenecarboxylic

Function
Preservative- Antimicrobial, Shelf Life Extender

Description

We use enzymes for fermentation and to make things like corn syrup, but we can't seem to keep them out of food when we want them gone. The problem is, foods like enzymes. It's only natural. Microorganisms such as bacteria, molds, and yeast produce them. Benzoic acid was discovered in the 16th century, as reported by none other than Nostradamus, but it wasn't until 1875 that German biochemist Ernst Leopold Salkowski discovered its antifungal talents by observing how well cloudberries, which contain benzoic acid, preserved food.

By 1909, scientific research proved that sodium benzoate (the salt of benzoic acid) could stop the microorganisms, and thus the enzymes, from growing. So sodium benzoate has become one of the most commonly used preservatives in history, despite the fact that no one knows how it actually works.

Here's what we do know: In an acidic food environment, sodium benzoate creates benzoic acid that accumulates on the cell walls of the microorganisms, which is a bad thing for the cell. Cellular activity shuts down.

Now widely used, sodium benzoate is found in foods (at no more than 0.1 percent concentration) such as carbonated and still beverages, syrups, margarine, olives, relishes, jellies, jams, pie fillings, prepared fruit and salads, stored vegetables, fruit juices, and, of course, pickles. It works just as well in pharmaceuticals, such as cough syrups, toiletries, and cosmetics. On top of all these uses, sodium benzoate figures in antifreeze, dye-making, photographic processing, and even whistling fireworks. It is also a rust inhibitor for iron. From time to time, sodium benzoate is prescribed as a medicine for reducing ammonia in the body, and in the past it has been an ingredient in topical antiseptics and antifungal cremes. It makes sense that what fights fungi in food would fight them in your toes.

One place you will not find it is in Coke and Pepsi. The companies dropped it as an ingredient due to consumer demand for natural ingredients. It seems that the toxic carcinogen benzene is created when combined with ascorbic acid [p. 12], heat, and sunlight. While the amount created was below tolerated levels (you get much more just from breathing urban air, pumping gas, or smoking), benzene is benzene and the outcry was convincing enough for the manufacturers to search for alternatives. The customer is always right, no matter what they say.

Sodium nitrite

NaNO$_2$

CFR number
172.175

E number
E250

CAS number
7632-00-0

Synonyms or siblings
Nitrous acid, Sodium salt

Function
Appeal-Color
Preservative-Antimicrobial, Antioxidant

Description

Sodium nitrite does two things extremely well: It's an extraordinarily good meat preservative, and it turns meat slightly pink, making it look fresh, longer. Back in the 1970s, it did a third thing well: It scared the living daylights out of people. Some newspaper articles pointed out that nitrites could combine with amines under certain conditions to create nitrosamines, some of which are carcinogenic. Even more incredible is that the scare was about nothing. Not only were facts and conditions elusive, the only thing scientists did seem to agree on at the time was that dangerous levels were found only in certain kinds of overcooked bacon, but nitrite levels are kept extremely low to reflect current consumer preferences and science.

Since time immemorial people have been curing meat and fish with salt, using it to dehydrate naturally present bacteria. Over time, people came to prefer sodium nitrate over sodium chloride, in part because of the pinkish color it gave the meat. Once sodium nitrate is in contact with food, it converts into sodium nitrite, making the two terms effectively interchangeable from a consumer's point of view.

The same thing happens when sodium nitrate is digested: our bodies convert it into sodium nitrite. And when do we digest sodium nitrate/nitrite? Whenever we eat fruits, grains, or leafy vegetables, such as spinach and celery. Depending on the soil and growing conditions, spinach may contain 500ppm to 1,900ppm of nitrate; lettuce, up to 1,700ppm; and radishes, up to 1,800ppm. The US Department of Agriculture allows no more than a minuscule 156ppm of nitrite to be added to cured meats, but typically, even at the common level of 120ppm, only about 10ppm remains after processing. Vegetables are the source of up 80 to 90 percent of our nitrite consumption.

Celery juice is the handy ingredient that allows food manufacturers to say "no added nitrates" on packaging. Instead of adding the chemical directly, they use celery juice to sidestep consumer wariness. Unfortunately, this often means that there are more, sometimes ten times more, nitrites than conventional foods might have.

Sodium nitrite is a standard product of the industrial chemical world, being the result of a series of basic reactions starting with a mixture of sodium hydroxide (lye), nitrogen dioxide, and nitric oxide. It is widely used as a precurser chemical in many other industrial processes.

Beyond its use in preventing botulism while preserving and coloring meat, sodium nitrite is used to make nitric acid, bright red and orange textile and leather dyes, a variety of organic compounds, metal coatings and rust inhibitors, and rubber chemicals. It's a medicine, too, useful as a vasodilator, bronchial dilator, intestinal relaxant, and even as an antidote for cyanide poisoning. Nitrite can also mitigate heart attack damage, speed wound healing, treat sickle cell anemia, and more. One thing it apparently does not do, however, is cause cancer. And here we were worried about hot dogs!

Sodium stearoyl lactylate (SSL)

$C_{24}H_{43}NaO_6$

CFR number
172.846

E number
E481

CAS number
25383-99-7

Synonyms or siblings
Sodium stearoyl-2-lactylate, Sodium stearyl-2-lactylate, Sodium 2-stearoyllactylate, Sodium stelate

Function
Process and Prep-
Emulsifier, Stabilizer

Description

Its name betrays its ingredients. Sodium stearoyl lactylate is made from sodium, stearic acid, and lactic acid [p. 98]. It's a fat that's also an emulsifier, making it incredibly useful in the processed-food business and in other popular products.

The two main ingredients come from plants on opposite sides of the world. Stearic acid is extracted from vegetable oil, most commonly palm oil harvested in Malaysia or Indonesia. Oil palm fruit is cooked, crushed, dehydrated, cleaned, and refined into an oil that can be shipped to the United States or Europe for further processing. Because it's already about half saturated fat, it doesn't need to be hydrogenated—palm oil is naturally thick and stable. The palm oil is zapped with superheated water, separating out the fatty acids in a fractionation tower, much like petroleum is fractionated to produce gasoline and kerosene. In this case, stearic acid is pumped around the corner to be hydrogenated. It thickens into something akin to candle wax while in transport and has to be heated again for handling later on. Stearic acid is usually part of various emulsifiers, such as polysorbate 60 [p. 126], that work together to provide the hallmark gooeyness for many shampoos and lotions.

Lactic acid [p. 98], the second main ingredient, is a clear, thin liquid made by fermentation and some simple chemical reactions from corn syrup, often in the Midwest. Both acids are shipped to an SSL manufacturer, most likely also in the Midwest.

There, the two acids are neutralized with a dash of an alkali, probably sodium carbonate or sodium hydroxide (lye). The three ingredients are cooked for about two hours in a massive kettle. The result is a thick, fatty syrup, which is then pumped over an ice-cold, slowly rotating, 10-foot-diameter, stainless-steel drum called a flaker. The cold turns the fat into a soapy wax that flakes off into containers for further breakup into a greasy, superfine, superuseful powder.

Sodium stearoyl lactylate might resemble the egg yolk in your home cooking, but in industrial cooking it's the ingredient that gives bakers the most bang for their buck because of its utility as both a fat and an emulsifier. It works in the dough, conditioning it for stability and strength, improving the protein, and making it easier to manipulate. Then SSL functions as an emulsifier, tightening and increasing the volume of the crumb, partly by helping to retain the gas created by the leavening. (Think Wonderbread as opposed to an airy French baguette). It "complexes" the starch, keeping bread softer longer. SSL isn't just limited to baked goods; it works as a whipping aid in cheese sauces, "crème" fillings, and puddings, while it also emulsifies coffee whiteners (artificial creamers) and low-fat margarines, among many other products.

Despite this heroic list of attributes, SSL remains a minor ingredient, usually less than 1 percent of a flour mixture, making it the crème de la crème of artificial food ingredients, so to speak.

dence that folic acid
defects in newborns
ent, millers went
r well before the
iance date—before
inted—and the FDA

With overwhelming e
could cut neural tube
by as much as 70 per
ahead and put it in fl
mandatory 1998 com
labels could even be
was happy to allow it.

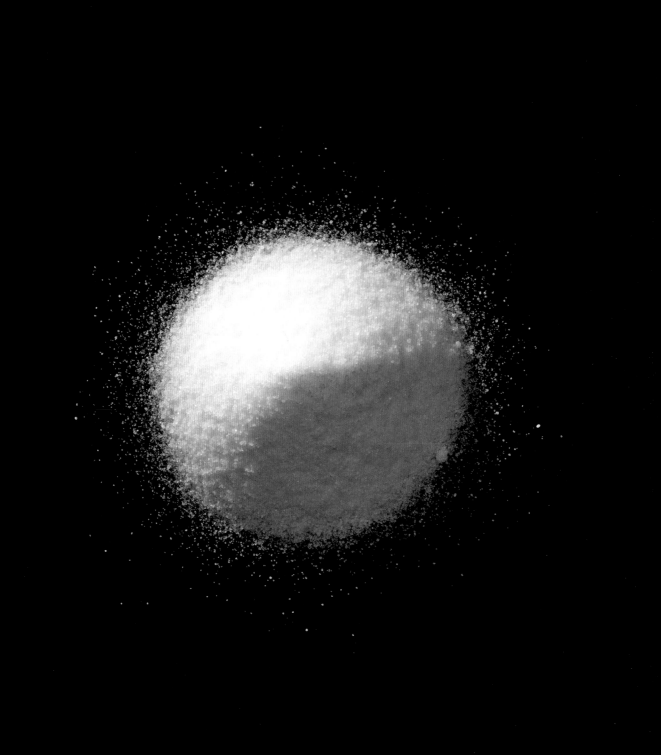

Sorbic acid

$C_6H_8O_2$

CFR number
182.3089

E number
E200

CAS number
110-44-1

Synonyms or siblings
2-Propenyl acrylic acid,
T-2, 4-Hexanedieonic acid

Function
Preservative-
Antimicrobial, Shelf Life Extender

Description

Nothing fights mold better than sorbic acid. It is extremely potent and needs to be present only in microscopic quantities in a given food product. Despite the fact that it's made in oil refineries and chemical plants in Germany, Japan, and China, it is actually food, a polyunsaturated fatty acid and thus a molecular cousin of common vegetable oils.

Sorbic acid manufacturing starts with natural gas that is cracked to make ethane and methane, which then are processed into various ingredients (such as ethylene oxide, methanol, and acetic acid) and reacted with various catalysts or components (such as palladium, carbon monoxide, and manganese) to create two different liquids, one clear and one light yellow. These are mixed together to create sugarlike crystals that can be ground into a powder for food use. Some is shipped out as pure sorbic acid, while some is mixed with the minerals potassium or calcium to make sister products calcium sorbate and potassium sorbate, all three the most popular preservatives in the world.

Sorbic acid is less toxic than table salt. The body metabolizes it like it would any other fatty acid, turning the sorbic acid into carbon dioxide and water.

Sorbic acid takes its name from the mountain ash tree genus Sorbus (which is a big improvement on its real name, trans trans 2, 4-hexadienoic acid). Since time immemorial, Old World mythology and folklore led humans to believe that the mountain ash or European rowan tree, with its bright orange berries, was magical. It was said to protect against malevolent beings and ward off evil influences. It turns out that the folklore was right.

Back in 1859, German chemist August Wilheim van Hoffman cooked up a batch of oil distilled from the juice of pressed, unripe mountain ash berries and managed to isolate sorbic acid for the first time. Apparently, the berries manufacture sorbic acid to protect themselves, resulting in some medicinal qualities. The vitamin C–loaded mountain ash berries are too astringent to eat on their own, but their fresh juice or tea has been used as a laxative and a cure for sore throats, inflamed tonsils, and hoarseness; their jam and infusions have been popular remedies for a host of other ailments, including diarrhea, indigestion, hemorrhoids, scurvy, and gout.

Sorbic acid was first synthesized only with the advent of the petrochemical industry, circa 1900. Still, despite the natural evidence, no one proved its antimicrobial properties until 1939, when two different scientists—one in Germany and one in the United States—discovered them simultaneously.

It's no surprise that at typical usage levels, you can't smell or taste sorbic acid, even in bland foods. It's strong stuff, used to the tune only of 0.2 percent of a batch of dough most of the time, and sometimes only 0.03 percent. That's barely three ounces per hundred pounds, a mere dusting. Call it the stealth additive.

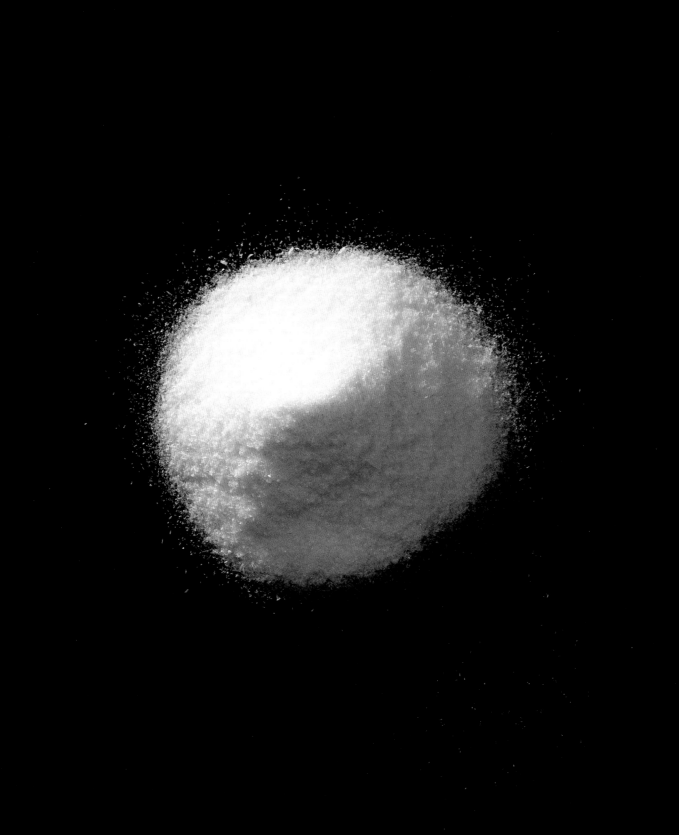

Soy Ingredients

A Mini-Primer on Soybean Processing

Oilseed processing—soy, corn, canola, and the like—is done on an enormous scale. One of the larger companies can handle more than 4 million bushels of soybeans each day at their plants around the world, and their 2013 profits from soybean processing alone topped $1.5 billion. The process is similar for soy alternatives such as cottonseed oil, sunflower oil, canola oil, and palm oil, but 80 percent to 90 percent of all vegetable oilseeds processed this way are soybeans.

Processing starts with the agricultural version of a giant sauna. First, cooked, cracked, and dehulled soybeans are tempered (gently heated to soften them) for a week or so in large silos. They then get crushed (into precisely eight pieces if all goes well) between the two rollers of the muscular cracking-roller machine. Next, the chunks (or chips, as some places call them) are shot into flaking rollers, which are about six feet long and have only a paper-thin gap in between them. The rollers crush the bean chunks into round, yellowish flakes the size of cornflakes. However, instead of being uneven (or frosted), these flakes are flat and naturally loaded with oil—about 20 percent of a soy flake is oil.

Most vegetable oils are not pressed like expensive expeller-pressed and cold-pressed vegetable oils. Instead, the oil is extracted from the flakes by soaking them with hexane—a toxic, explosive petroleum product that is a very effective solvent (also a gasoline and glue component)—in a building-size tower. Hazardous solvents

may seem incompatible with food, but they have a welcome habit of disappearing when boiled off after causing the oil to separate from the flakes—they're volatile, after all. Warm water is mixed into the gummy, crude oil and immediately absorbed by any sticky gum residue, and the oil is then centrifuged out, leaving a brownish-black sludge and the flakes. The degummed oil then goes on to become refined soybean (cooking) oil and partially hydrogenated vegetable oil (shortening), the leftover sludge is processed into lecithin, and the defatted flakes—the solid residue— go on to become soy protein isolate, all described in the following pages.

Partially hydrogenated vegetable oil

CFR number
101.4

CAS number
68334-28-1
8016-70-4
(Partially hydrogenated soybean oil)

Synonyms or siblings
Partially hydrogenated (soybean, cottonseed, etc.,) oil, Hydrogenated vegetable oil, Hydrogenated (soybean, cottonseed, etc.,) oil

Function
Preservative-Leavening Agent

$C_{57}H_{98}O_6$
1, 2, 3-linoleoyl glycerol

Description

Before it can become shortening, degummed soybean oil is bleached with sodium hydroxide (caustic soda) and mixed with a claylike product that removes chlorophyll and the remaining carotenoid pigments that color it yellow. Heating to about 500 degrees Fahrenheit for up to an hour under a near-full vacuum—about 1 millimeter of pressure—and blasting with steam vaporizes any remaining free fatty acids, off-flavors, and moisture. Finally, after this extensive refining, the oil has the light, bland color and pourable liquid consistency that we like for salad dressings, baking, and frying. Some will also end up in an amazing array of industrial goods, including (after more processing, of course) alkyd paints—the current version of "oil" paint—rubber, caulk, adhesive tape, leather softeners, and diesel fuel.

But oil is not the goal here—a soft solid that resembles butter or lard is. The next step, hydrogenation, creates a semisolid oil that behaves like butter but costs a whole lot less, contains as much as 50 percent less saturated fat, and is so stable that it can be stored without refrigeration. All you have to do is force some hydrogen molecules into it. Oil to which only some hydrogen has been added is called "partially hydrogenated."

Shortening is a blend of partially and sometimes fully hydrogenated oils and up to 80 percent unhydrogenated, liquid oil along with emulsifiers and antioxidants. This is all in an effort to reduce the famously unwanted trans fats created by partial hydrogenation of regular soybean oil (no trans fats are in unhydrogenated, liquid oil or in fully hydrogenated, solid oils).

Hydrogenation was invented in 1905 in France (despite being the land of fabulous butter), and led directly to Crisco in 1911—then made from cottonseed oil (its name, selected in a Procter & Gamble employee contest, is a near-acronym derived from Crystalized Cottonseed Oil). Crisco was one of the very first artificial food ingredients. It is now trans fat-free.

The basic difference between shortening and butter is that shortening's crystalline structure and high melting point creates air pockets in dough during baking, ensuring an airy cake or a flaky crust. Shortening offers a particular advantage over butter or other fats: it makes cakes tender by coating the flour proteins with oil, keeping them from absorbing moisture, and shortening the gluten strands that would otherwise make the baked good dense. Hence, the otherwise mysterious name.

Soy lecithin

CFR number
184.1400

E number
E322

CAS number
8002-43-5

Synonyms or siblings
Phosphatidylcholine,
Soybean lecithin

Function
Process and Prep-
Emulsifier, Stabilizer,
Thickening Agent

$C_{44}H_{80}NO_8P$
Phosphatidylcholine

Description

Lecithin is ubiquitous in the natural world. At least trace amounts of it are found in almost all living cells. Once considered a waste product from soybean oil processing, it is now one of the most common emulsifiers in food processing.

French scientist Maurice Gobley was the first to identify lecithin, back in 1805. He found that egg yolks are about 30 percent lecithin. Gobley named his discovery after the Greek word for egg yolk, *lekithos*. European industrial soybean oil processing did not start in earnest until around 1908. In 1920, more than a century after Gobley's discovery, German food scientists identified the stinky, dark-brown waste sludge left over from that processing as lecithin.

Lecithin additive is refined from that sludge, which is slightly heated to remove water until it resembles molasses. The processors further refine it into clear liquid or mix it with a little oil to make granules and powders. Since it's much easier to extract from soy than from eggs, soy lecithin became one of the very first industrially produced emulsifiers used by commercial bakeries.

Lecithin often works in concert with its synthetic counterparts, mono- and diglyceriedes [p. 104] and polysorbate 60 [p. 126]. Lecithin seems indispensible to bakers, as it improves dough handling, moisture retention, texture, volume, browning, and shelf life, all while improving the effectiveness of shortening and reducing the need for expensive and perishable egg yolks. It also serves the delectable purpose of keeping chocolate smooth and reducing "bloom," that white haze of fat that sometimes forms on the surface of chocolate confections. Lecithin is also used to smooth out and bind ingredients in ice cream, chewing gum, and peanut butter as well as whipped topping, processed cheese, and dry beverage mixes. This is merely a short list of its emulsifying abilities: wetting agent, instantizer (helping things dissolve), release agent (in cooking spray), antidusting agent, and more. On top of that versatility, it's high in polyunsaturates, cholesterol-free, and totally safe to eat.

Not surprisingly, lecithin has many nonfood uses as well. It emulsifies paint pigments, water-based printing inks, plastics, and the coatings on videotapes; it works as a skin softener in cosmetics, helps oil penetrate leather, and plays minor roles in paper coating, waxes, adhesives, lubricants, and explosives. This useful goo has risen to far surpass its former tag as a waste product.

Soy protein isolate (SPI)

CFR number
101.82

CAS number
9010-10-0

Synonyms or siblings
Textured soy protein,
Soy protein concentrate,
Soybean protein

Function
Nutrient-Fortifying Agent

Description

One of the latest, most high-tech ingredients in food scientists' arsenal is soy protein isolate, or SPI. After soybeans are crushed in flakes and the oil removed, the mushy, high-protein flakes go to a specialized plant where they are dumped, 20,000 gallons at a time, into vats of warm water and lye, lime, ammonia, or tribasic phosphate. The result is a soggy mess akin to a watery milkshake. After about an hour of gentle agitation, the proteins and sugars are dissolved and the protein can be extracted in centrifuges the size of cars in a building not unlike an airplane hangar.

The second step is like making tofu. A bit of acid, usually hydrochloric, is added to the moist mix in a 4,000- or 5,000-gallon tank to trigger a curdling reaction. Curiously for this well-known dairy alternative, the proteins (the soy protein isolate) are called "curds" and the watery effluent, "whey."

Finally, the third step is to dry the soy protein, now about a 90 percent to 95 percent concentration, into the desired shapes and sizes: finely powdered for drinks and baking; coarse grains for the creation of soy-based meat substitutes. Three tons of soy flakes make one ton of isolated protein, explaining why it costs about five to seven times as much as soy flour (the same soy flakes used for SPI are also the source for soy flour, soy grits, and textured vegetable protein).

As fabulous as isolated soy protein is for food manufacturers, that doesn't mean it can't be used by other industries. For instance, SPI helps bind the clay coating on certain kinds of cardboard, including the exterior of cereal boxes. You can thank soy science for the colorfully printed tigers, toucans, leprechauns, and other beloved food mascots on your cornflakes.

Stevia

E number
E960

CAS number
91722-21-3

Synonyms or siblings
Steviol Glycosides, Candy leaf, Sugar leaf, Sweetleaf

Function
Appeal-Sweetener

$C_{38}H_{60}O_{18}$
Stevioside

Description

Stevia is a simple additive with a complicated history. The green leaves of this native South American plant have been used for hundreds of years to sweeten common foods and drinks, like maté tea. It is the No. 1 sugar substitute in Japan, where consumers go through hundreds of tons of it every year. It has been studied in countless ways, and the European Union has allowed stevia since 2011. However, the Food and Drug Administration, having banned it for years, finally allowed only a highly purified stevia extract, rebaudioside A, in 2008.

Regulations are the least of the inconsistencies associated with stevia. Some studies show that stevia or some of its components can reduce insulin levels, thus offering benefits to people who have diabetes or other sugar issues. Other studies show no benefits of stevia use. Either way, consumer studies show no harmful side effects. In 1995, it was accepted as a dietary supplement but not as a food additive, meaning that the government found it simultaneously safe and unsafe. That's complex.

What is clear is that the stevia plant derivatives—steviol glycocides, especially rebaudioside A—create a zero-calorie, zero-carbohydrate, zero–glycemic index sugar alternative. One manufacturing method uses a water process to extract what's essential, while others use alcohol solvents that are evaporated after processing (Cargill and Coca-Cola call theirs by its common name, rebiana). Commercial products tend to be blended with water, alcohol, or vegetable fiber to make various forms of this sugar substitute. The glycocide compounds are about 300 to 400 times as sweet as sugar, so the product is generally mixed with fillers to create a sugar equivalent that masks its slightly bitter and lingering aftertaste; some processors can get around this and sell a pure product.

Stevia is not new. Spanish botanist and physician Petrus Jacobus Stevus (Pedro Jaime Esteve in Spanish; he signed his work in Latin) first studied stevia plants back in the early 1500s, and established the Latinized name for the plant, Stevia rebaudiana). Amazonian tribes in what is now Paraguay and Brazil used it for centuries before that for both cooking and medicinal uses. In 1931, two French chemists isolated the active chemicals behind the sweetness, and in the early 1970s a Japanese artificial-sweetener company, finding its older products banned or shunned, developed both cultivation and processing techniques for stevia that avoid its bitter aftertaste. It's a bittersweet story.

Sucralose

$C_{12}H_{19}Cl_3O_8$

CFR number
172.831

E number
E955

CAS number
56038-13-2

Synonyms or siblings
Trichlorogalactosucrose (TGS), Trichlorosucrose, Acucar light, Aspasvit, Splenda

Function
Appeal-Sweetener

Description

Sweeteners seem to be the main artificial ingredient people love to hate. Entire Web sites are devoted to their discredit (though often such sites are easily discredited themselves either through lack of credentials or lack of credibility). That said, it is true that there were serious scares surrounding the first artificial sweeteners. Sucralose is the newest sugar substitute in the United States, and its 1991 Canadian approval, like its 1998 and 1999 US approvals (and that of 80 other countries), was preceded by twenty years of much-touted testing results. These showed that sucralose is not only safe but also without effect on the body (i.e., also insulin and glucose) or on the environment. Obviously, long-term human studies have yet to conclude; without these, sucralose critics will continue to cast their doubt.

The discovery of sucralose is a famous, well-circulated anecdote. In 1976, a researcher at Queen Elizabeth College in London was directed by a scientist (who was trying to create new insecticides) to test a chlorinated sugar compound. The researcher misunderstood, hearing "taste" instead of "test," and found incredible sweetness (straight sucralose is 600 times sweeter than sugar). Further analysis led to the realization that the compound is heat-stable and water-soluble (though not fat-soluble), and that the company behind these studies should patent the stuff. It is sold to consumers under the name Splenda, which is a mix of sucralose and maltodextrin [p. 176,60] blended to equal sugar in volume when used at home. Industrial food and beverage companies, who buy it uncut, use it in more than 4,500 products.

Sucralose is artificial, but its manufacture does start with sugar. The largest plant is on an island in Singapore, owned by sucralose originator Splenda, while the No. 2 plant is in China. The rather complex fabrication involves replacing the hydrogen-oxygen atom groups of the sugar molecule with chlorine atoms. The chlorine, usually in the form of phosphorous oxychloride (but also hydrogen chlorine, thionyl chloride, and benzyltriethlyammonium chloride), reacts in such a way that the resulting molecule, trichlorogalactosucrose, is totally stable and nonreactive; the chlorine in the compound stays put. It is so stable that it is not affected by heat nor, for the most part, by human digestion. Most of the sucralose you consume is passed out of your body having only reacted with the sugar receptors on your taste buds.

Sucralose alone is without calories, but the bulking agents such as maltodextrin add small amounts of calories and carbohydrates. The amounts are below the level needed to report their presence when found in sweeteners used only a teaspoon at a time, but they can add up when using more. There's no free lunch.

Sugar

$C_{12}H_{22}O_{11}$

CFR number
184.1854

CAS number
57-50-1

Synonyms or siblings
Sucrose, Cane juice, Crystalline fructose, Dextran

Function
Appeal-Sweetener

Description

Providing sweetness is really only sugar's side job. Especially in baked goods, sugar carries flavor, provides color, fosters tenderness, creates an even crumb, and retains some moisture in order to improve shelf life. Other so-called sweeteners can neither share nor simulate sugar's versatility and importance in baked goods. Without a lot of sugar, cake would essentially just be bread.

Sugar starts its work before even entering the oven, during the creaming stage of any dessert recipe. The irregular surfaces of sugar crystals trap air in small pockets that will later expand in the oven, yielding flaky pastry. In the mixing stage, sugar tenderizes cake by combining with all the protein it can and absorbing water that would otherwise help build protein and its elasticity. Sugar also helps stabilize beaten egg white foam (via linking proteins, and only if not overbeaten) in confections like soufflés, sponge cake, and meringue.

Sugar caramelizes on the surface when its temperature reaches over 300 degrees Fahrenheit, browned surfaces not only taste, smell, and look good; they also retain a bit more moisture than the inner part of the baked item due to their relative denseness. Finally, sugar steals water from bacterial cells, further preventing spoilage, which is why sugar is such a well-known preservative. Just think of jams, jellies, all kinds of "preserves" made with sugar, or of containers of honey that last for years.

Multitalented sugar tenderizes and improves the appearance of canned fruit, delays discoloration on the surface of frozen fruits, and enhances flavors in all kinds of desserts, especially ice cream. Sugar balances sour, bitter, and hot flavors in spicy dishes; balances acidic foods, like tomatoes and vinegar, or sour and bitter tastes in rubs and brines; enhances mouthfeel in drinks and sauces; and strengthens fiber in fruits and vegetables during cooking. Sugar makes dry baked goods, like cookies, crisp. Of course, let's not forget that in both nature and in processed foods, sugar makes everything more palatable. We're hardwired to like sweet treats.

Cane processing plants wash the shredded and crushed stalks in hot water, which dissolves the sugar in order to produce cane juice. Impurities are removed by adding lime (calcium hydroxide). The juice is filtered and evaporated into

Taurine

$C_2H_7NO_3S$

CFR number
573.980

CAS number
107-35-7

Synonyms or siblings
Tauric acid, L-Taurine, Tauphon

Function
Nutrient-Fortifying Agent

Description

Even though the substance from which taurine was first extracted has not been used for production since around 1860, its appeal to young men chugging taurine-laced energy drinks has not diminished. The word "taurine" is obviously rooted in the Latin word "taurus," for bull; it's an appealing link for consumers seeking to take their manliness to new heights and for marketers of consumer goods trying to exploit that. Alas, taurine is no longer extracted from bull semen or even from cow livers these days, though that was the traditional Chinese medicine source for years—now it's produced by dozens of Chinese chemical companies. Even the Red Bull Energy Drink Web site says that taurine, which is actually called 2-aminoethanesulfonic acid, L-Taurine, is now mostly produced synthetically. You can still find it naturally: your body makes it as a big part of your bile.

The most common chemical path uses ethylene oxide, a basic petrochemical, as its raw material. This is reacted with another food additive, aqueous sodium bisulfite. A less common but similar path mixes an organic chemical, aziridine, with sulfurous acid. Not terribly romantic, but efficient.

Ah, well, as many of those youthful searchers of manliness might say, whatever. At least energy drinks do provide lots of water and sugar [p. 178], plus caffeine [p. 26].

Taurine is an amino acid that helps regulate water and mineral salt levels in your blood. It is a building block for all the other amino acids. It courses through your central nervous system, muscles, eyes, gallbladder, and brain. On a cellular level, taurine plays a role in the movement of potassium, sodium, calcium, magnesium, and zinc. In traditional medicine, taurine has been used to treat congestive heart failure, ischemic heart disease, arrhythmias, epilepsy, hypertension, cystic fibrosis, diabetes, liver failure, and more. Some suggest it as treatment for colds, fevers, tonsillitis, and even drug poisoning. A common ingredient in pet food—an industry that uses half the global taurine production—it is an especially important nutrient for cats because they cannot synthesize it themselves. Taurine may have antioxidant properties.

Dietary sources include meat, fish, seaweed, and breast milk. Not quite as stimulating as an all-night reveler or hardworking student might want, but there it is. Combined with caffeine, taurine might possibly improve mental abilities, but no one knows for sure. One thing is certain, though, and that is that excess taurine is excreted by your kidneys.

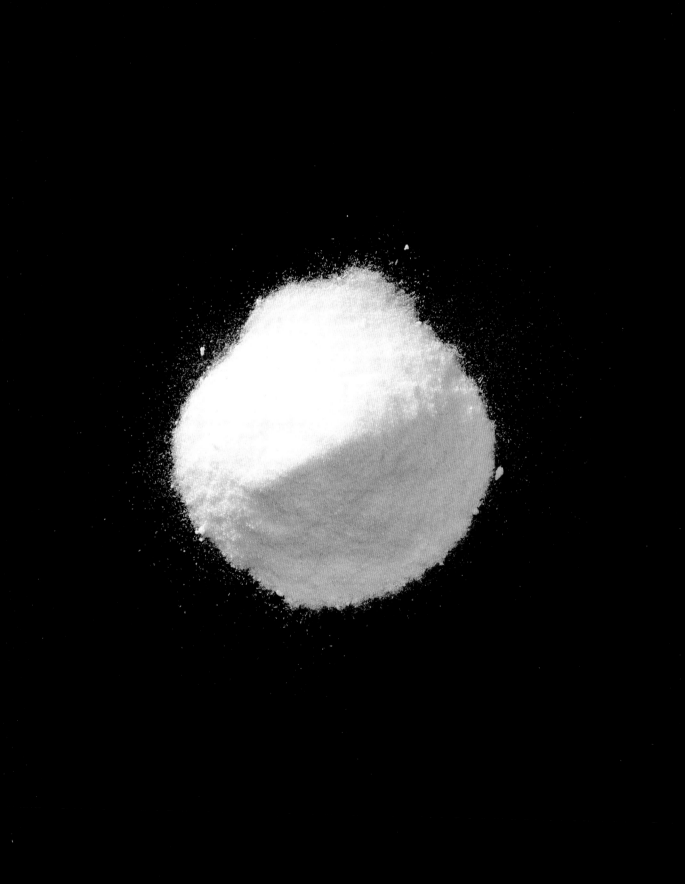

Tertiary-butylhydroquinone (TBHQ)

$C_{10}H_{14}O_2$

CFR number
172.185

E number
E319

CAS number
1948-33-0

Synonyms or siblings
Tert-butylhydroquinone, tBHQ

Function
Preservative-Antioxidant

Description

It doesn't smell, but it is used in perfumes. It has a name that scares a lot of people, but it is hugely popular. It is used in extremely low doses as an antioxidant, but that also makes it a good corrosion inhibitor in biodiesel fuel. It helps slow the evaporation of varnishes and lacquers. And most of us can't pronounce it: tertiary-butylhydroquinone. Luckily, it can be easily abbreviated to TBHQ (or tBHQ).

One reason people started to get scared by TBHQ is that someone said it was made from butane, even though it isn't. It is a petroleum product made in labs that depend on oil refineries for their raw materials (as do so many organic chemical plants) and it is made almost entirely in India or China. In high doses it can be toxic or even fatal. How high of a dose are we talking about? One scientist figured you'd need to eat at least 312.5 McDonald's Chicken McNuggets in just one day in order to get sick, and you'd need five times that to die. It doesn't accumulate in your body, which is why you'd have to consume all those nuggets in one day; your body excretes antioxidants such as TBHQ quite efficiently, as it should. Despite these facts, some consumers still feel it should be avoided.

TBHQ is limited to 0.02 percent of the fat and oil content of any food product, and levels are similarly limited for cosmetics. Since it fights rancidity particularly well, it is often used along with BHA and BHT [p. 24] or tocopherol to make sure everything stays fresh. It works best in vegetable oils, animal fats (lard), essential oils (hence the cosmetics track record), and nuts. This is why it is found in baked goods, confections, snack foods, fried foods, cereals, and various emulsifiers and flavorings.

Sometimes it is dissolved into the wax coating on packaging, mixed with the oil as the food product is made, sprayed onto the food, mixed into the seasoning, or cooked right into the food product. In short, TBHQ is found anywhere and everywhere.

Thiamine mononitrate

$C_{12}H_{17}N_5O_4S$

CFR number
184.1878

CAS number
532-43-4

Synonyms or siblings
Vitamin B1, Mononitrate de thiamine, Nitrate de thiamine, Thiamine nitrate, Thiamin

Function
Nutrient-Enriching Agent

Description

Thiamine mononitrate, or vitamin B1, was the very first vitamin to be discovered. In the late 1800s, Dutch scientist Christiaan Eijkman was working in Indonesia when he realized that only people whose diet included white rice, from which the brown husk—the rice bran—had been removed, suffered from the awful, nerve-damaging disease beriberi. ("Beriberi," a Sinhalese word meaning "I cannot, I cannot," became the name of the disease because the victim suffers extreme stiffness of the lower limbs, pain, and even paralysis, and is too ill to do anything.) It turns out that while thiamine is found in small quantities in a variety of foods—meaning you shouldn't lack it if you eat a balanced diet—it is found in, and readily absorbed from, brown rice.

By isolating the factor that was essential to health in this one case, Eijkman concluded that certain chemicals in food were essential to health in general, laying the groundwork for the discovery of all vitamins a few years later (and a Nobel Prize in Medicine in 1929). As it turns out, we need B1 to help our bodies make fuel; it plays an important role in converting carbohydrates into energy. Of course, it plays other roles as well, such as keeping most of our cardiac, muscular, and nervous systems in good shape. That's why, since 1942, it's been one of the vitamins included in enriched white flour.

Getting thiamine ready for this, its most common use, is an extensive process. The manufacturing varies from company to company and is an especially guarded secret. What we do know is that thiamine mononitrate is usually synthesized not from the brown rice that led to its discovery but from basic petrochemicals derived from that trusty old food source, coal tar.

Thiamine chemicals are finished with about fifteen steps that may include, depending on the company, such appetizing processes as oxidization with corrosive-strength hydrogen peroxide and active carbon; reactions with ammonium nitrate, ammonium carbonate, and nitric acid (to form an edible salt); and washing with alcohol. The resulting liquid is dried into crystals and sieved into a fine powder than can be easily mixed in minute quantities into enriched white flour.

In 2005, the world's largest fine-chemical company, the German firm BASF, started a cooperative venture with the Tianjin Zhongjin Pharmaceutical Company. A couple of hours north of Beijing, it's now the world's largest B1 plant. Each year, BASF expects to produce 3,000 tons of a material that is used by a fraction of an ounce at a time. With all that thiamine at the ready, it's no wonder that we don't get beriberi anymore.

Titanium dioxide

TiO_2

CFR number
73.575

E number
E171

CAS number
13463-67-7

Synonyms or siblings
Titania, Rutile, Anatase, Brookite

Function
*Appeal-*Color
Process and Prep-
Thickening Agent

$O = Ti = O$

Description

When it comes to the color white, nothing touches titanium dioxide—nothing. It is the whitest of whites, whiter than almost anything else found in the world. People refer to its ultra whiteness by name, "titanium white." It reflects light so well that it is used in most high-SPF sunblocks. It literally shines as a pigment in the industrial world, in white paints and other coatings, inks, and paper; nothing else is as bright. In fact, food coloring is the least of its uses. Most of it is used in industrial products, from the highest-quality white paint to plastics to cosmetics. Scientists found that it even has electronic qualities, leading to its use in computer screens, and hydrophilic aspects that are employed to make self-cleaning glass and antifogging glass coatings.

When it is used in food, it is used sparingly. Most popular in white dairy products—think skim milk—it's also a very common component of candy, especially sugarcoated confections. It lightens the color of a wide range of foods, from soups to nuts (literally!), including dried vegetables and mustards. It also disperses well and serves as an insoluable thickener. It adds whiteness to things that go in your mouth but are not food, too, such as cigarette papers, toothpaste (of course), and chewing gum.

Titanium dioxide starts as a highly processed mineral dug out of mines—usually anatase, a hard rock. The crude TiO_2 is often converted into titanium tetrachloride first. After soaking it in various bases and acids to separate it out, the TiO_2 crystals break down into particles so small that some are considered nanoparticles. Although it's considered a safe additive, the use of nanoparticles in cosmetics, especially sunblocks, has come under suspicion. This may be because there is a possibility that the ultrasmall particles can be absorbed by the skin. More studies are needed to assuage the critics, though the authorities have found it to be safe.

Above all, titanium dioxide is common. Among the top 50 minerals mined in the world, it accounts for 70 percent, globally, of all pigments made. Although the statistics are not clear, China claims to export 200 million tons of it each year. That's a lot of nanoparticles.

Xanthan gum

$(C_{37}H_{56}O_{30})n$

CFR number
172.695

E number
E415

CAS number
11138-66-2

Synonyms or siblings
Bacterial polysaccharide, Corn sugar gum, Goma xantana

Function
Process and Prep—Emulsifier, Stabilizer, Thickening Agent

Description

One of the most popular gums in use today is so hygroscopic—meaning it quickly absorbs and is transformed by water—that it must be stored in specially sealed containers. Classified broadly, xanthan gum is a thickener, viscosity-increasing agent, suspending agent, stabilizer, and emulsifier. It is a staple item in the food scientist's tool bag for keeping ingredients from separating while serving as a fat substitute (as in salad dressing), or to keep ice crystals from forming while creating a smooth texture (as in ice cream).

Xanthan gum is used for other gooey, thick concoctions, including medicines (such as cough syrup), toothpaste, and cosmetics (such as sunscreen). It also absorbs oil, making it popular in the oil-spill cleanup industry. Other nonfood applications include coagulating any watery liquid, including such unlikely pairs as drilling mud and fake blood.

In the natural world, this gum is the mucouslike slime that many fungi and algae secrete to keep themselves moist; however, it is one of the newest food ingredients on the market, having been approved only in 1968. Thanks to a US Department of Agriculture research project geared toward finding useful corn products, xanthan gum was discovered in the 1950s by a chemical researcher named Allene R. Jeanes who specialized in carbohydrates.

Though considered a natural product, xanthan gum is made in ingredient factories constructed next door to giant corn syrup factories so that a dextrose/glucose solution can be delivered directly to fermenting tanks for use as bacterial food. (Some manufacturers use dairy whey [p. 182], a cheese-making by-product full of lactose, as the sugar source.) A bacterium, Xanthomonas campestris—the source of this gum's unusual name, as well as the black rot on broccoli and cauliflower—is mixed into the syrup. The bacteria feed and multiply for one to four days before being precipitated out of the fermenter by a dose of isopropyl alcohol. The result is a starchy goo that is dried, powdered, and sold as a popular, concentrated food additive.

Processed-food companies are not the only ones that cook with xanthan gum. Packaged as a fairly common grocery store item, it is an excellent addition to gluten-free baking recipes. Its rivals include guar gum and locust bean gum. With an astounding seven grams of fiber per tablespoon (29 percent Daily Value!), it is a concentrated fiber source (though always used in small quantities because it can cause bloating) and can help people who have constipation or difficulty swallowing. Xanthan gum is so potent that most recipes call for using a concentration of less than 0.5 percent. Any more and it might gum up the works.

Part 2

25 Processed Food Products

Amy's Burrito Especial

01 Organic black beans
02 Organic wheat flour
03 Filtered water
04 Organic onions
05 Organic rice
06 Organic tomato puree
07 Organic bell peppers
08 Expeller-pressed high-oleic safflower
09 and/or Sunflower oil

Cheddar and Monterey Jack Cheeses:
10 Pasteurized milk
11 Culture
12 Salt
13 Annatto (color)
14 Enzymes (without animal enzymes or animal rennet)

15 Sea salt
16 Olives
17 Organic green chiles
18 Cilantro
19 Organic garlic
20 Spices
21 Organic sweet rice flour

Campbell's Chunky Classic Chicken Noodle Soup

01 Chicken stock
02 Chicken meat
03 Carrots

Enriched Egg Noodles:
04 Wheat flour
05 Egg whites
06 Eggs
07 Niacin
08 Ferrous sulfate
09 Thiamine mononitrate
10 Riboflavin
11 Folic acid

12 Celery

Contains 2% or less of:
13 Modified food starch
14 Salt
15 Chicken fat
16 Potassium chloride
17 Soy protein concentrate
18 Yeast extract
19 Sugar
20 Dehydrated mechanically separated chicken
21 Dehydrated onions
22 Cooked chicken skins
23 Sodium phosphate
24 Flavoring
25 Spices
26 Beta-carotene for color
27 Dehydrated vegetable broth
28 Disodium guanylate
29 Disodium inosinate
30 Dehydrated chicken
31 Egg yolks
32 Soy lecithin

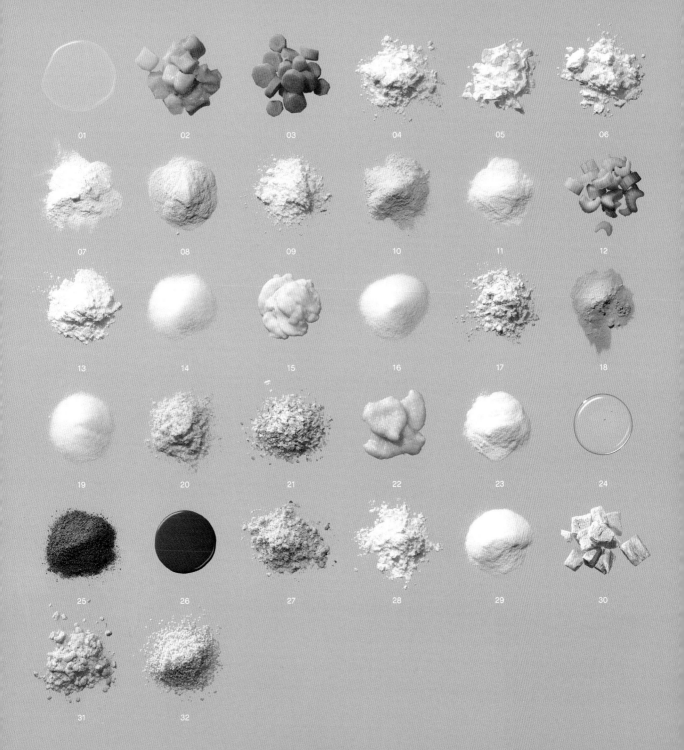

Doritos Cool Ranch Flavored Tortilla Chips

01 Corn

Vegetable Oil:
02 Corn oil
03 Canola oil
04 and/or Sunflower Oil

05 Maltodextrin (made from corn)
06 Salt
07 Tomato powder
08 Cornstarch
09 Lactose
10 Whey
11 Skim milk
12 Corn syrup solids
13 Onion powder
14 Sugar
15 Garlic powder
16 Monosodium glutamate

Cheddar Cheese:
17 Milk
18 Cheese cultures
19 Salt
20 Enzymes

21 Dextrose
22 Malic acid
23 Buttermilk
24 Natural flavor
25 Artificial flavor
26 Sodium acetate

Artificial color including:
27 Red No. 40
28 Blue No. 1
29 Yellow No. 5

30 Sodium caseinate
31 Spices
32 Citric acid
33 Disodium inosinate
34 Disodium guanylate

General Mills Raisin Nut Bran

01 Whole grain wheat
02 Sugar
03 Raisins
04 Almonds
05 Corn bran
06 Corn syrup
07 Brown sugar syrup
08 Partially hydrogenated cottonseed and/or soybean oil
09 Salt
10 Glycerin
11 Molasses
12 Cornstarch
13 Soy lecithin
14 Trisodium phosphate
15 Natural flavor
16 Artificial flavor
17 BHT added to preserve freshness

Vitamins and Minerals:

18 Calcium carbonate
19 Zinc
20 and Iron (mineral nutrients)

21 Vitamin C (sodium ascorbate)
22 A B Vitamin (niacinamide)
23 Vitamin B_6 (pyridoxine hydrochloride)
24 Vitamin B_2 (riboflavin)
25 Vitamin B_1 (thiamine mononitrate)
26 A B Vitamin (folic acid)
27 Vitamin B_{12}

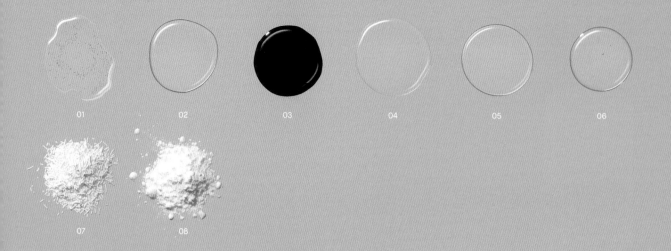

Dr Pepper

01 Carbonated water
02 High-fructose corn syrup
03 Caramel color
04 Phosphoric acid
05 Natural flavor
06 Artificial flavor
07 Sodium benzoate (preservative)
08 Caffeine

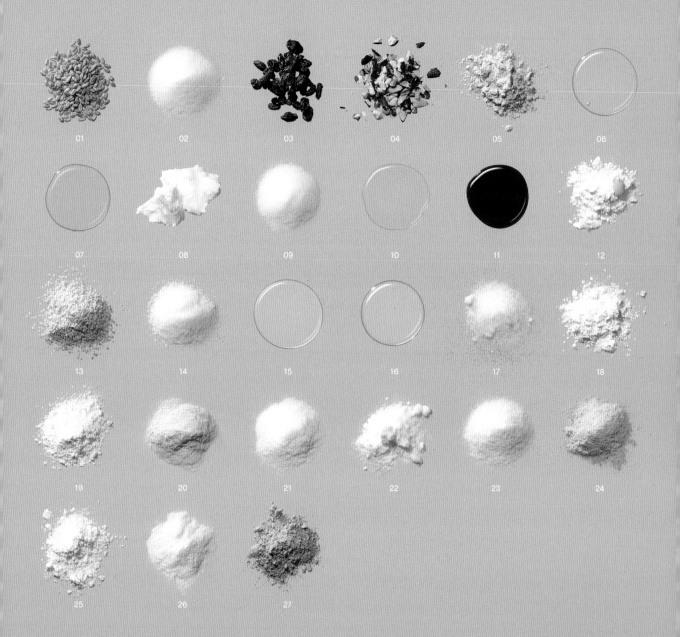

Hebrew National Beef Franks

01 Beef
02 Water

Contains 2% or less of:
03 Salt
04 Spice
05 Sodium lactate
06 Paprika
07 Hydrolyzed soy protein
08 Garlic powder
09 Sodium diacetate
10 Sodium erythorbate
11 Flavoring
12 Sodium nitrite

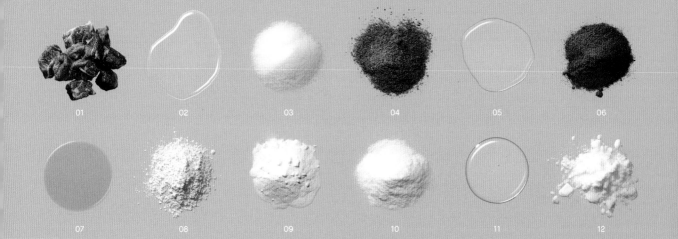

Heinz Tomato Ketchup

01 Tomato concentrate from red ripe tomatoes
02 Distilled vinegar
03 High-fructose corn syrup
04 Corn syrup
05 Salt
06 Spice
07 Onion powder
08 Natural flavoring

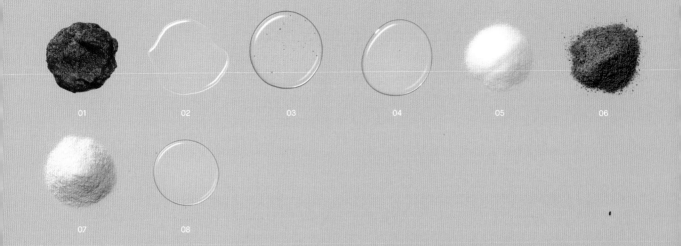

Hidden Valley The Original Ranch Light Dressing

01 Water
02 Vegetable oil (soybean
03 and/or canola)
04 Maltodextrin
05 Buttermilk
06 Sugar
07 Salt

Contains 2% or less of:
08 Spices
09 Dried garlic
10 Dried onion
11 Natural flavors (soy)
12 Egg yolk
13 Modified food starch
14 Phosphoric acid
15 Vinegar
16 Artificial flavor
17 Disodium phosphate
18 Xanthan gum
19 Monosodium glutamate
20 Artifical color
21 Disodium inosinate
22 Disodium guanylate
23 Sorbic acid and
24 Calcium EDTA as preservatives

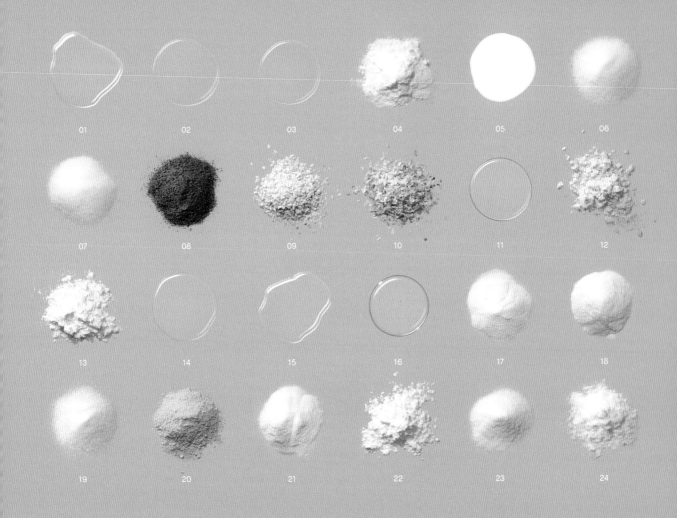

Hostess Twinkies

Enriched Bleached Flour:
01 Flour
02 Reduced iron
B Vitamins:
03 Niacin
04 Thiamine mononitrate (B_1)
05 Riboflavin (B_2)
06 Folic Acid

07 Water
08 Sugar
09 Corn syrup
10 High-fructose corn syrup

Partially Hydrogenated Vegetable and/or Animal Shortening:
11 Soybean oil
12 Cottonseed oil
13 and/or Canola oil
14 Beef fat

15 Whole eggs
16 Dextrose

Contains 2% or less of:
17 Soy lecithin

Leavenings:
18 Sodium acid pyrophosphate
19 Baking soda
20 Cornstarch
21 Monocalcium phosphate

22 Modified cornstarch
23 Glucose
24 Whey
25 Glycerin
26 Soybean oil
27 Salt
28 Monoglycerides
29 Diglycerides
30 Polysorbate 60
31 Cornstarch
32 Sodium stearoyl lactylate
33 Natural flavor
34 Artificial flavor
35 Sorbic acid (to retain freshness)
36 Potassium sorbate
37 Xanthan gum
38 Cellulose gum
39 Enzyme
40 Wheat flour
41 Yellow No. 5
42 Red No. 40

Klondike Reese's Ice Cream Bars

Light Ice Cream:
01 Nonfat milk
02 Sugar

Reese's Peanut Butter:
03 Peanuts
04 Sugar
05 Peanut oil
06 Dextrose
07 Salt
08 TBHQ (preserves freshness)

09 Corn syrup
10 Milkfat
11 Whey
12 Maltodextrin
13 Propylene glycol monoesters
14 Cellulose gum
15 Monoglycerides
16 Diglycerides
17 Cellulose gel
18 Locust bean gum
19 Guar gum
20 Polysorbate 80
21 Carrageenan
22 Caramel color
23 Vitamin A palmitate
24 Annatto (for color)

Reese's Peanut Butter Cup Pieces:
Milk Chocolate:
25 Sugar
26 Cocoa butter
27 Chocolate
28 Nonfat milk
29 Milkfat
30 Corn syrup
31 Soy lecithin
32 PGPR (polyglycerol polyricinoleate)

33 Peanuts
34 Sugar
35 Dextrose
36 Salt

Milk Chocolate Flavored Coating:
37 Coconut oil
38 Sugar
39 Chocolate liquor (processed with alkali)
40 Soy lecithin
41 Artificial flavor
42 Salt

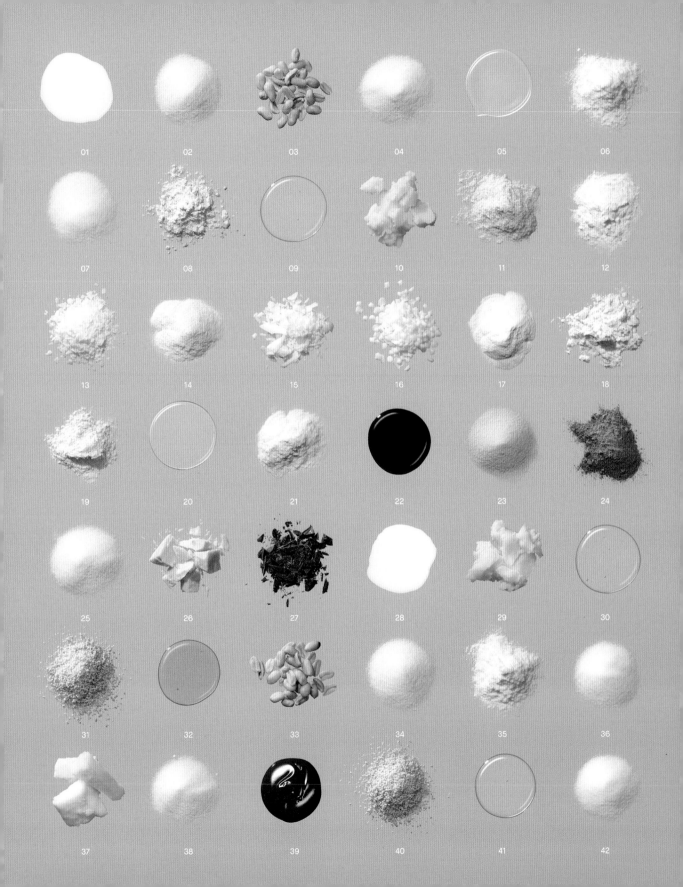

Kraft Cool Whip Original

01 Water

Hydrogenated Vegetable Oil:
02 Coconut oil
03 Palm kernel oil

04 High-fructose corn syrup
05 Corn syrup
06 Skim milk

Contains 2% or less of:
07 Light cream
08 Sodium caseinate
09 Natural flavor
10 Artifical flavor
11 Xanthan gum
12 Guar gum
13 Polysorbate 60
14 Sorbitan monostearate
15 Sodium polyphosphate
16 Beta-carotene (color)

Kraft Singles - American Skim Milk Fat Free

01 Whey

Skim Milk Cheese:
02 Skim milk
03 Cheese culture
04 Salt
05 Enzymes

06 Water
07 Milk protein concentrate

Contains 2% or less of:
08 Sodium phosphate
09 Dried corn syrup
10 Salt
11 Milk
12 Whey protein concentrate
13 Sodium hydroxide
14 Lactic acid
15 Sodium citrate
16 Sorbic acid as a preservative
17 Artificial color
18 Cellulose gum
19 Artificial flavor
20 Carrageenan
21 Disodium inosinate
22 Cheese culture
23 Vitamin A palmitate
24 Enzymes
25 Apocarotenal (color)
26 Annatto (color)

McDonald's Chicken McNuggets

01 White boneless chicken
02 Water
03 Food starch-modified
04 Salt

Seasoning:
05 Autolyzed yeast
06 Salt
07 Wheat starch
08 Natural flavor (botanical source)
09 Safflower oil
10 Dextrose
11 Citric acid

12 Sodium phosphates
13 Natural flavor (botanical source)

Battered and breaded with:
14 Water
Enriched flour:
15 Bleached wheat flour
16 Niacin
17 Reduced iron
18 Thiamine mononitrate
19 Riboflavin
20 Folic acid

21 Yellow corn flour
22 Bleached wheat flour
23 Food starch-modified
24 Salt

Leavening:
25 Baking soda
26 Sodium acid pyrophosphate
27 Sodium aluminum phosphate
28 Monocalcium phosphate
29 Calcium lactate

30 Spices
31 Wheat starch
32 Dextrose
33 Corn starch

Prepared in Vegetable Oil:
34 Canola oil
35 Corn oil
36 Soybean oil
37 Hydrogenated soybean oil
38 with TBHQ
39 and Citric Acid to preserve freshness of the oil
40 and Dimethylpolysiloxane to reduce oil splatter when cooking

MorningStar Farms Original Sausage Patties

Textured Vegetable Protein:
01 Wheat gluten
02 Soy protein concentrate
03 Soy protein isolate
04 Water for hydration

05 Egg whites
06 Corn oil
07 Sodium caseinate
08 Modified tapioca starch

Contains 2% or less of:
09 Lactose
10 Soybean oil
11 Hydrolyzed vegetable protein (wheat gluten, corn gluten, soy protein)
12 Autolyzed yeast extract
13 Spices
14 Natural flavor
15 Artificial flavor

Sodium phosphates:
16 Tripolyphosphate
17 Tetrapyrophosphate
18 Hexametaphosphate
19 Monophosphate

20 Salt
21 Disodium inosinate
22 Caramel color
23 Cellulose gum
24 Whey powder
25 Modified cornstarch
26 Maltodextrin
27 Potassium chloride
28 Dextrose
29 Onion powder
30 Disodium guanylate

Vitamins and Minerals:
31 Niacinamide
32 Iron (ferrous sulfate)
33 Thiamine mononitrate (vitamin B_1)
34 Pyridoxine hydrochloride (vitamin B_6)
35 Riboflavin (vitamin B_2)
36 Vitamin B_{12}

37 Succinic acid
38 Ascorbic acid
39 Lactic acid
40 Brewer's yeast
41 Torula yeast
42 Soy lecithin

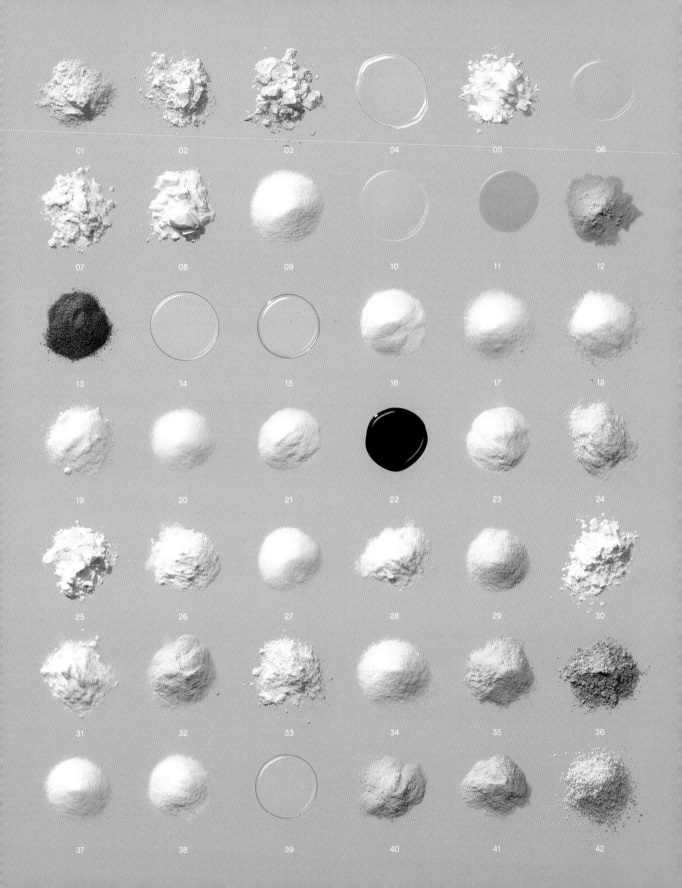

Nabisco Wheat Thins

01 Whole grain wheat flour
02 Soybean oil
03 Sugar
04 Cornstarch
05 Malt syrup (from corn and barley)
06 Salt
07 Refiner's syrup

Leavening:
08 Calcium phosphate
09 and/or baking soda

Vegetable color:
10 Turmeric oleoresin
11 Annatto extract

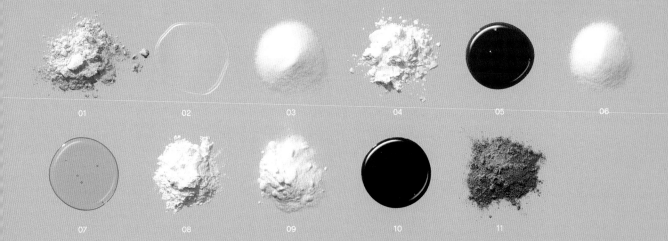

Naked Green Machine 100% Juice Smoothie

01 Apple juice
02 Mango puree
03 Pineapple juice
04 Banana puree
05 Kiwi puree
06 Spirulina
07 Natural flavors
08 Alfalfa
09 Broccoli
10 Spinach
11 Barley grass
12 Wheat grass
13 Parsley
14 Ginger root
15 Kale
16 Odorless garlic

Nestlé Coffee-Mate Fat Free The Original Coffee Creamer

01 Corn syrup solids

Hydrogenated Vegetable Oil:
02 Coconut oil
03 and/or Palm kernel oil
04 and/or Soybean oil

05 Sugar
06 Sodium caseinate (a milk derivative)
07 Dipotassium phosphate

Contains 2% or less of:
Color added
08 Monoglycerides
09 Diglycerides
10 Sodium aluminosilicate
11 Natural flavor
12 Artificial flavor
13 Salt
14 Annatto color

Ocean Spray Cran-Grape Juice Drink

01 Filtered water

Grape juice:
02 Water
03 Grape juice concentrate

04 Cane or beet sugar

Cranberry juice:
05 Water
06 Cranberry juice concentrate

07 Fumaric acid
08 Natural flavors
09 Vegetable concentrate for color
10 Sodium citrate
11 Ascorbic acid (vitamin C)
12 Citric acid

PowerBar Performance Energy Bar Oatmeal Raisin

Dual Source Energy Blend:
01 Cane invert syrup
02 Maltodextrin
03 Fructose
04 Dextrose

05 Raisins
06 Soy protein isolate
07 Whole oats (contains wheat)
08 Oat bran

Brown Rice Crisps:
09 Rice bran
10 Rice flour
11 Rosemary extract

12 Brown rice flour
13 High oleic canola oil

Contains 2% or less of:
14 Calcium phosphate
15 Vegetable glycerin
16 Sugar
17 Salt
18 Potassium phosphate
19 Cinnamon
20 Ascorbic acid (vitamin C)
21 Natural flavor
22 Partially defatted peanut flour
23 Nonfat milk
24 Ground almonds
25 Allspice
26 Nutmeg
27 Ferrous fumarate (iron)
28 Pyridoxine hydrochloride (vitamin B_6)
29 Thiamine hydrochloride (vitamin B_1)
30 Riboflavin (vitamin B_2)

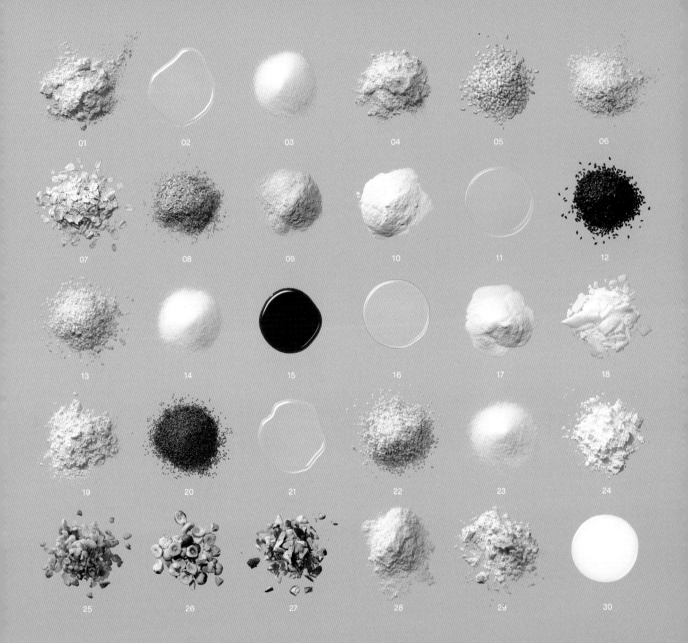

Oroweat Healthy Multi-Grain Bread

01 Whole wheat flour
02 Water
03 Sugar
04 Wheat gluten
05 Brown rice
06 Cornmeal
07 Oats
08 Wheat bran
09 Yeast
10 Cellulose fiber
11 Soybean oil
12 Black sesame seeds
13 White sesame seeds
14 Salt
15 Molasses
16 Datem
17 Calcium propionate (preservative)
18 Monoglycerides
19 Calcium sulfate
20 Poppy seeds
21 Grain vinegar
22 Soy lecithin
23 Citric acid
24 Calcium carbonate

Nuts:
25 Walnuts
26 Hazelnuts (filberts)
27 and/or Almonds

28 Whey
29 Soy flour
30 Nonfat milk

Red Bull Energy Drink

01 Carbonated water
02 Sucrose
03 Glucose
04 Citric acid
05 Taurine
06 Sodium bicarbonate
07 Magnesium carbonate
08 Caffeine
09 Niacinamide
10 Calcium pantothenate
11 Pyridoxine HCl
12 Vitamin B_{12}
13 Natural flavor
14 Artificial flavor

Colors including:
15 Red No. 40
16 Blue No. 1
17 Yellow No. 5

Quaker Instant Oatmeal Strawberries & Cream

01 Whole grain rolled oats
02 Sugar

Flavored and Colored Fruit Pieces:
03 Dehydrated apples (treated with sodium sulfite to promote color retention)
04 Artificial strawberry flavor
05 Citric acid
06 Red No. 40

Creaming Agent:
07 Maltodextrin
08 Partially hydrogenated soybean oil
09 Whey
10 Sodium caseinate

11 Salt
12 Calcium carbonate
13 Guar gum
14 Artificial flavor
15 Citric acid
16 Niacinamide
17 Reduced iron
18 Vitamin A palmitate
19 Pyridoxine hydrochloride
20 Riboflavin
21 Thiamine mononitrate
22 Folic acid

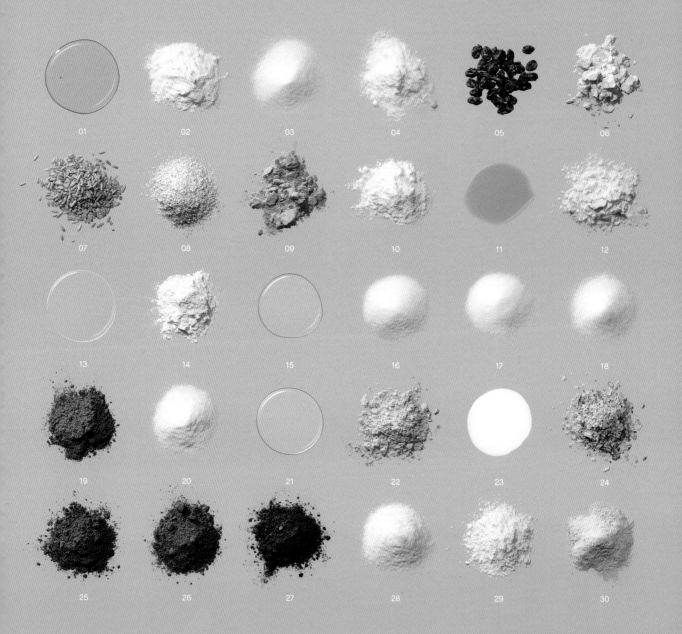

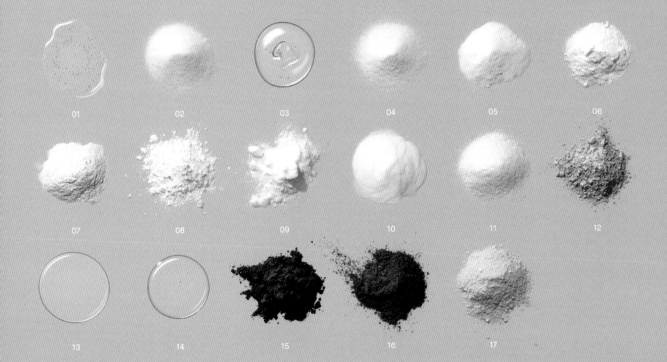

Snickers Bar

Milk Chocolate:
01 Sugar
02 Cocoa butter
03 Chocolate
04 Skim milk
05 Lactose
06 Milkfat
07 Soy lecithin
08 Artificial flavor

09 Peanuts
10 Corn syrup
11 Sugar
12 Milkfat
13 Soy lecithin
14 Partially hydrogenated soybean oil
15 Lactose
16 Salt
17 Egg whites
18 Chocolate
19 Artificial flavor

Trident Perfect Peppermint Sugar Free Gum

01 Sorbitol
02 Gum base
03 Mannitol
04 Xylitol
05 Glycerin
06 Natural flavoring
07 Artificial flavoring

Contains 2% or less of:
08 Acesulfame potassium
09 Aspartame
10 BHT (to maintain freshness)

Glossary

Additive, food additive
Nonfood intentionally added directly or indirectly to food product as part of processing or packaging.

Antioxidant
Molecule that blocks oxidation (see below). May prevent or delay some cell damage by highly reactive chemicals called free radicals. Found in many common fruits and vegetables.

Binder
Additive that keeps various ingredients from separating for improved mouthfeel.

CAS number
A numeral assigned to every chemical substance (more than 80 million at present) by the Chemical Abstracts Service, a division of the American Chemical Society.

C.I. number
Identification number issued by Colour Index International to ensure consistent color by manufacturers around the world. Also abbreviated CI.

Dough conditioner
Additive that stabilizes (or "improves") dough by strengthening the gluten or other means so that it stands up to industrial manufacturing and provides a more consistent product.

E number
Identifying code for food additives approved for use within the European Union and Switzerland and used by many other countries. Issued by the European Food Safety Authority on the basis of the Codex Alimentarious committee. "E" stands for "Europe."

Emulsifier
Additive that binds fat and water to prevent separation, makes smoother mixing of ingredients, controls crystallization, and helps ingredients dissolve.

Enzyme
Natural chemical catalyst, usually a protein, which speeds up specific biological processes, notably digestion.

FDA
US Food and Drug Administration, charged with regulating food and drugs. The FDA created and regulates the "Nutrition Facts" ingredients label.

Foodstuff
Food with nutritive value meant to be eaten, whether vegetable, animal, or fungal.

Fractionated
Result of fractionation, in which a substance (especially oil) is heated to the point where its components separate out. Most often done in a tower where the various components, or fractions, leave the original substance at various heights.

Gel
Additive that produces viscosity through a gelling effect for uniform texture and improved mouthfeel.

GRAS
Acronym for Generally Recognized as Safe, an FDA regulatory category for a partial list of additives that, according to qualified experts and based on substantial history, has been adequately shown to be safe (and is not otherwise listed).

Humectant
Additive that binds and holds moisture.

Hydrogenation
The process of pumping hydrogen gas into oil at high pressure and temperature with the help of a nickel or similar catalyst in order to turn the oil into a semisolid (partially hydrogenated) or waxy solid (fully hydrogenated). Partial hydrogenation creates unwanted trans fatty acids in place of desirable unsaturated fatty acids.

Hydrolyzed
The effect of hydrolysis, in which chemical bonds of a substance are broken by the chemical or mechanical addition of superheated, superpressurized water.

Hygroscopic
Readily binds and holds moisture; substance is slightly transformed by doing so.

3. Make Food More Appealing

Enhance Flavor: Supplement, magnify, or modify the original taste and/or aroma of food without imparting a new flavor.

Add Color: Give desired, appetizing, or characteristic color of food.

Add Flavor: Heighten natural flavor; restore flavors lost in processing.

Add Sweetness: Make the aroma or taste of food more agreeable.

Stimulate: Raise nervous activity.

4. Preserve Product Quality and/or Freshness

Prevent Oxidation: Antioxidants slow or prevent changes in color, flavor, or texture and delay rancidity.

Prevent Food Spoilage: Antimicrobials prevent food spoilage from bacteria, molds, fungi, or yeast.

Extend Shelf Life: Maintain freshness.

Vlasic Ovals Hamburger Dill Chips

01 Cucumbers
02 Water
03 Distilled vinegar
04 Salt
05 High-fructose corn syrup
06 Calcium chloride
07 Natural flavors
08 Polysorbate 80
09 Yellow No. 5

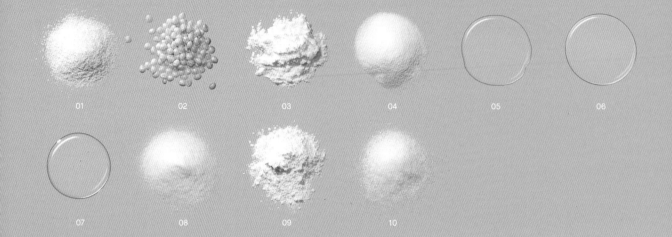

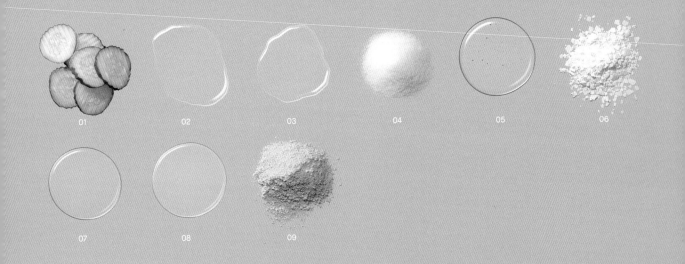

Food Additive Functions

1. Maintain or Improve Nutritional Quality

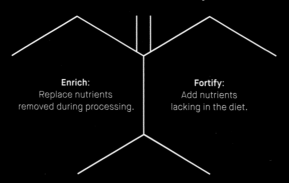

Enrich: Replace nutrients removed during processing.

Fortify: Add nutrients lacking in the diet.

2. Aid in Processing or Preparation

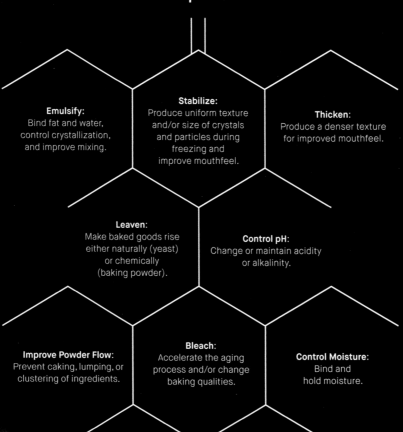

Emulsify: Bind fat and water, control crystallization, and improve mixing.

Stabilize: Produce uniform texture and/or size of crystals and particles during freezing and improve mouthfeel.

Thicken: Produce a denser texture for improved mouthfeel.

Leaven: Make baked goods rise either naturally (yeast) or chemically (baking powder).

Control pH: Change or maintain acidity or alkalinity.

Improve Powder Flow: Prevent caking, lumping, or clustering of ingredients.

Bleach: Accelerate the aging process and/or change baking qualities.

Control Moisture: Bind and hold moisture.

Natural ingredient
One without color or artificial substances added.

Organic
Food grown or made without additives, pesticides, antibiotics, or added hormones; organic products with at least 70 percent organic ingredients.

Oxidation
Technically defined as the loss of an electron; generally defined as the interaction between oxygen and other molecules. Food oxidizes by turning brown and spoiling.

Phytochemical
Compounds made by plants and that affect human health. They are non-nutritive and protect the plants.

PPM
Parts per million, a common chemical measurement of concentration.

Texturizer
Additive that produces uniform texture and improves consistency for improved mouthfeel.

Whole food
Food in its natural state without additives; unprocessed food.

A Word About Chemical Formulas
Chemical notations indicating commonly used molecular formulas of many ingredients described are included for informal reference only. Notations and names may vary among vendors and manufacturers. Also, some different additives have the same molecular formula but different molecular structures, making them different compounds. Readers should not assume that the molecular formulas shown are definitive.

As a quick refresher, please note that "C" is for carbon, "H" is for hydrogen, and "O" is for oxygen—the most common elements of the compounds in this book. The other element symbols come from throughout the periodic table.

Acknowledgments

This project would not have happened were it not for a few people.

Thanks to my agent, Lindsay Edgecombe at Levine Greenberg Rostan Literary Agency, for thinking this should be a book. Thank you, Michael Szczerban for agreeing with Lindsay and getting the ball rolling.

Thank you, Steve Ettlinger, for your enthusiasm, energy, sense of humor, and talented writing. I'm lucky to have had the opportunity to collaborate with you.

I want to thank our publisher Regan Arts and the editors involved with this project; Lynne Ciccaglione, Sydney Tanigawa, and Brittany Dulac. In particular, Sydney, I want to thank you not just for your editing talents but your patience as well.

I'm especially grateful to Vanessa Chu and her talented hand for styling this project. I also appreciate her patience in executing many hours of research, along with many more hours spent organizing our library of additives. Thank you Jamie Antonioli for your help in completing additional research, along with sourcing many of the hundreds of ingredients photographed for this book. Also, great masking!

Thanks go to Tom Crabtree and Jerod Rivera at Manual design for making this book look amazing.

I'm lucky to have access to the retouching talents of Alex Katz and crew at blinklab. Their contribution to this book's photography is invaluable.

Thanks, Mom, for consistently reminding me that my diet could use improvement and that it's important to think about what I consume. I promise to stop drinking coffee some day, just not yet.

Thanks go to Daniel Patterson and Kevin Crafts for spurring the idea to do something artful with the elements of our diet and for simply showing me that food can be amazing.

And finally, thanks to my wife, Amy, and my sons, Harley and Theo. I'm proud that my son Harley is even more obsessed by Xanthan gum than I.

— Dwight Eschliman

Some captions in this book were adapted from full chapters containing more information on those ingredients in my earlier book, *Twinkie, Deconstructed*. This material is used with full permission of the publisher. I am very grateful to my editors and publishers at Hudson Street Press/Plume for their support.

More importantly, I wish to thank my children, Chelsea and Dylan, for the original inspiration to complete this book as well as for their continuing inspiration as energetic food cooks, shoppers, and growers. Thanks also to my wife, Gusty, for her continued support and constant enthusiasm for my creative projects.

I am extremely happy to have met Dwight Eschliman, an unusually talented and, above all, really nice guy. His dedication to a personal art project deserves all the support it can get.

Thanks to my savvy, cool editors at Regan Arts, Lynne Ciccaglione, Sydney Tanigawa, and Brittany Dulac, and our publisher, Judith Regan, for making this happen.

I am grateful for the information and assistance I have received from government, academic, corporate, and consumer publications and Web sites, as well as leading members of the food science profession and academic community. A special thanks goes to Joe Formanek of Ajinomoto North America, Inc., Jeff Greaves of Food Ingredient Solutions, LLC, Marion Nestle of New York University, Richard D. Ludescher, PhD of Rutgers University, and consultant and author Lazlo Somogyi. Some experts have reviewed portions of this text, but I am solely responsible for its accuracy.

— Steve Ettlinger

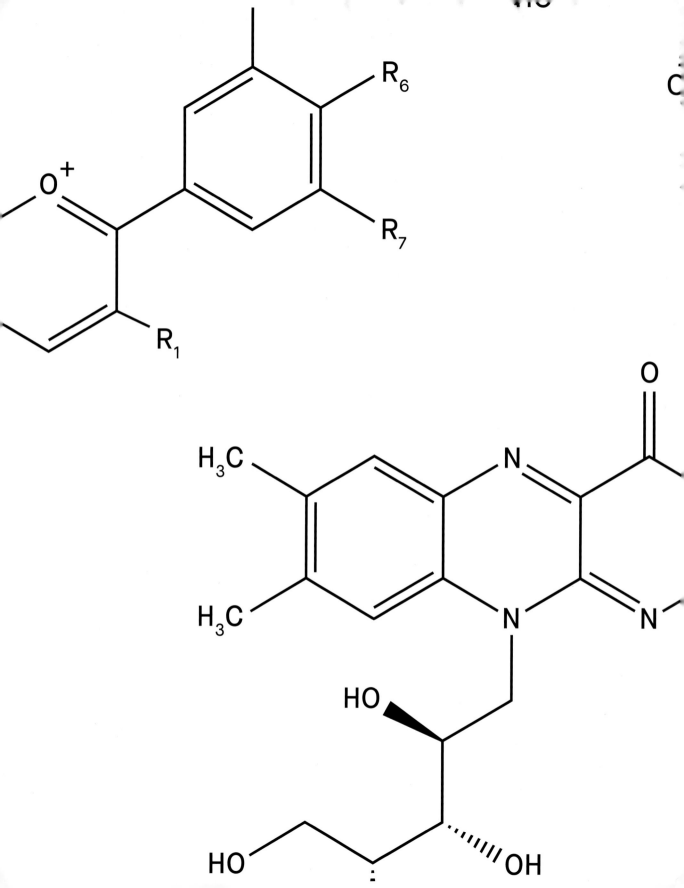

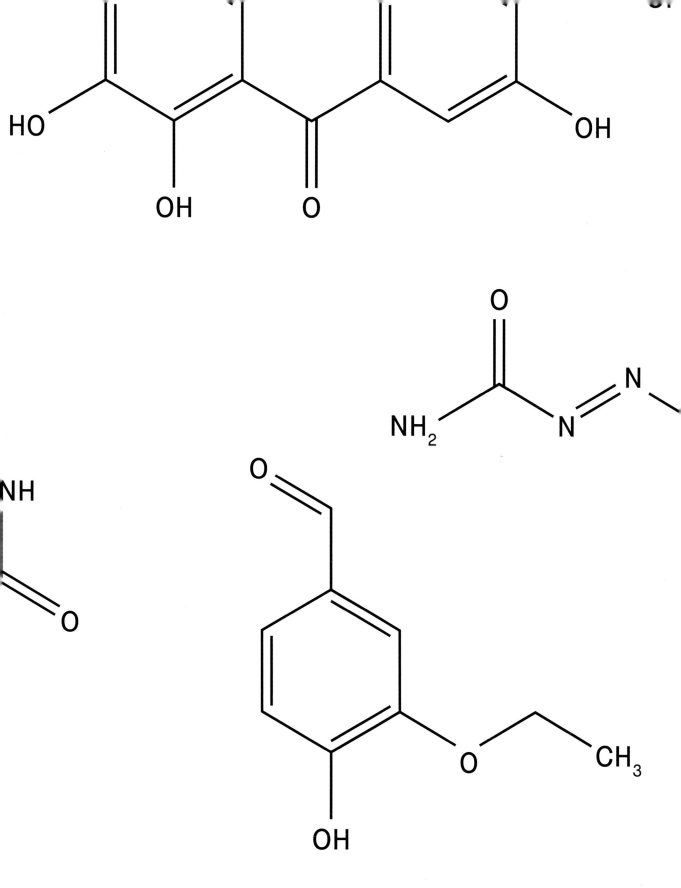

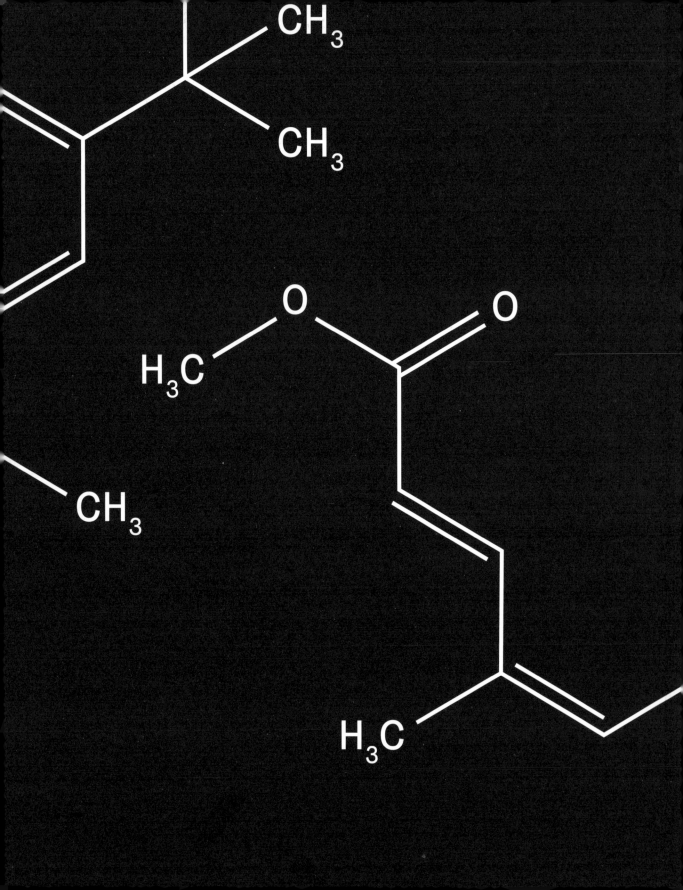